Imported Oil and U.S. National Security

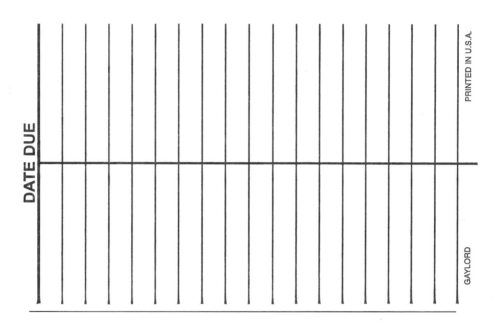

DATE DUE

PRINTED IN U.S.A.

GAYLORD

Keith Crane, Andreas Goldthau, Michael Toman, Thomas Light,
Stuart E. Johnson, Alireza Nader, Angel Rabasa, Harun Dogo

Sponsored by the Institute for 21st Century Energy
U.S. Chamber of Commerce

INFRASTRUCTURE, SAFETY, AND ENVIRONMENT and
NATIONAL SECURITY RESEARCH DIVISION

The research described in this report was sponsored by the Institute for 21st Century Energy, which is affiliated with the U.S. Chamber of Commerce, and co-conducted by the Environment, Energy, and Economic Development Program within RAND Infrastructure, Safety, and Environment and the International Security and Defense Policy Center of the RAND National Security Research Division.

Library of Congress Cataloging-in-Publication Data

Imported oil and U.S. national security / Keith Crane ... [et al.].
    p. cm.
    Includes bibliographical references.
    ISBN 978-0-8330-4700-7 (pbk. : alk. paper)
    1. Petroleum industry and trade—United States. 2. Petroleum industry and trade—Government policy—United States. 3. National security—United States. I. Crane, Keith, 1953-

    HD9566.I528 2009
    382'.422820973—dc22

                               2009010050

AP Photo/Kamran Jebreili

Published 2009 by the RAND Corporation
1776 Main Street, P.O. Box 2138, Santa Monica, CA 90407-2138
1200 South Hayes Street, Arlington, VA 22202-5050
4570 Fifth Avenue, Suite 600, Pittsburgh, PA 15213-2665
RAND URL: http://www.rand.org/
To order RAND documents or to obtain additional information, contact
Distribution Services: Telephone: (310) 451-7002;
Fax: (310) 451-6915; Email: order@rand.org

Preface

About This Document

The purpose of this study is to critically evaluate commonly suggested links between imported oil and U.S. national security and to assess the costs and benefits of potential policies for reducing U.S. consumption and imports of oil and to alleviate national security challenges linked to imported oil. We wrote this monograph to help policy-makers and the public evaluate the potential risks associated with importing oil and the extent to which policies might effectively reduce those risks.

The study was sponsored by the Institute for 21st Century Energy, which is affiliated with the U.S. Chamber of Commerce, and co-conducted by the Environment, Energy, and Economic Development Program (EEED) within RAND Infrastructure, Safety, and Environment (ISE) and the International Security and Defense Policy Center (ISDP) of the RAND National Security Research Division (NSRD). As with all RAND research, RAND maintains full editorial control over the content and conclusions of its reports. In this monograph, we draw on the expertise of several independent experts who reviewed the technical basis, findings, and conclusions to ensure their accuracy and balance.

The report is part of RAND research on energy issues. Recent publications include *Impacts on U.S. Energy Expenditures and Greenhouse-Gas Emissions of Increasing Renewable-Energy Use* (Toman, Griffin, and Lempert, 2008), *Producing Liquid Fuels from Coal: Prospects and Policy Issues* (Bartis, Camm, and Ortiz, 2008), and *Oil Shale Development in the United States: Prospects and Policy Issues* (Bartis et al., 2005).

The views in this publication do not necessarily reflect the opinions or policy positions of the sponsor, the U.S. Chamber of Commerce's Institute for 21st Century Energy.

The RAND Environment, Energy, and Economic Development Program

This research was co-conducted under the auspices of the EEED within ISE. The mission of ISE is to improve the development, operation, use, and protection of society's essential physical assets and natural resources and to enhance the related social aspects of safety and security of individuals in transit and in their workplaces and communities. The EEED research portfolio addresses environmental quality and regulation, energy resources and systems, water resources and systems, climate, natural hazards and disasters, and economic development—both domestically and internationally. EEED research is conducted for government, foundations, and the private sector.

Questions or comments about this monograph should be sent to the project leader, Keith Crane (Keith_Crane@rand.org). Information about EEED is available online (http://www.rand.org/ise/environ). Inquiries about EEED projects should be sent to the following address:

Keith Crane, Director
Environment, Energy, and Economic Development Program, ISE
RAND Corporation
1200 South Hayes Street
Arlington, VA 22202-5050
703-413-1100, x5520
Keith_Crane@rand.org

International Security and Defense Policy Center

This research was co-conducted within the ISDP of NSRD. NSRD conducts research and analysis for the Office of the Secretary of Defense, the Joint Staff, the Unified Combatant Commands, the defense agencies, the Department of the Navy, the Marine Corps, the U.S. Coast Guard, the U.S. Intelligence Community, allied foreign governments, and foundations.

For more information on ISDP, contact the Director, James Dobbins. He can be reached by email at James_Dobbins@rand.org; by phone at 703-413-1100, extension 5134; or by mail at the RAND Corporation, 1200 S. Hayes Street, Arlington, VA 22202. More information about RAND is available at www.rand.org.

Contents

Figures

Tables

Summary

Linkages Between Imported Oil and U.S. National Security

The United States consumes 25 percent of all the oil produced in the world, yet the United States accounts for only 10 percent of world oil production. In 2007, on a net basis, the United States imported 58 percent of what it consumes. This monograph critically evaluates commonly suggested links between these imports of oil and U.S. national security and assesses the costs and benefits of potential policies to alleviate challenges to U.S. national security linked to imported oil. We focus on the following areas of concern:

- economic
  - the potential for an abrupt fall in supply and the concomitant surge in the world market price of oil to disrupt U.S. economic activity to the point of precipitating an economic recession
  - damage to critical nodes in the U.S. supply chain for refined oil products that could induce short-run local shortages or, if extensive enough, national shortfalls in refined oil products, resulting in a reduction in U.S. economic output
  - large increases in payments by U.S. consumers of oil due to shifts in oil prices because of deliberate reductions in supply by major exporters
- political
  - the potential of major oil exporters to manipulate exports to influence other countries in ways inimical to U.S. interests
  - the potential for competition for oil supplies to exacerbate international tensions or disrupt international oil markets
  - the effect of higher revenues from oil exports on the ability of "rogue" oil exporters, such as Venezuela and Iran, to thwart U.S. policy goals
  - the potential role of oil-export revenues in supporting terrorist groups
- military: the additional costs to the U.S. defense budget of forces fielded to protect the supply and transit of oil from the Persian Gulf.

Economic Linkages

The gap between U.S. production and consumption is so large that eliminating it would entail extraordinarily costly changes to patterns of consumption and production of fuels. Moreover, even if total U.S. imports were cut sharply, the price of oil in the United States would still be determined by global, not national, shifts in supply and demand. A large, extended reduction in the global supply of oil would trigger a sharp rise in the price of oil and lead to a sharp fall in economic output in the United States, no matter how much or how little oil the United States imports.

The U.S. domestic supply chain for petroleum products is robust. Accelerated repairs of breakdowns, increased imports of refined oil products, and alternative domestic sources of supply make it highly unlikely that interruptions in domestic supplies could severely disrupt the U.S. economy.

Because the United States is a net importer of oil, when oil prices fall, as they did in the second half of 2008, the United States benefits from an improvement in its terms of trade, as consumers of refined oil products pay less for oil. Substantial reductions in U.S. consumption of oil or increases in domestic production of oil or oil substitutes would lower oil prices. A decline in oil prices may benefit the United States economically, if the cost of producing additional domestic fuel does not exceed the cost of importing oil and the economic costs of reducing oil consumption do not exceed the benefit of reduced oil costs. Lower oil prices would also benefit the U.S. military, which is a large consumer of refined oil products.

Political Linkages

Embargoes on exports of oil (and natural gas) have been unsuccessful in changing policies of targeted nations. As long as oil is a globally traded commodity, exporters cannot successfully target specific countries because importers can purchase alternative supplies on the global market.

Sales of oil below market prices or through grants have been more effective than embargoes at altering the behavior of targeted nations, but this limited support tends to last only as long as the favorable treatment.

Higher oil-export revenues have enhanced the ability for rogue states, such as Iran and Venezuela, to pursue policies contrary to U.S. interests.

The importance of donations from individuals and charities in oil-rich Middle Eastern states for financing al Qaeda and its affiliates has declined as terrorist groups have increasingly turned to crime to finance their attacks. Moreover, the costs of perpetrating a terrorist attack are so small ($15,000 to $500,000) that even a substantial fall in Middle Eastern oil revenues would not affect al Qaeda's ability to raise sufficient funds to finance its operations.

Incremental Costs of U.S. Forces to Secure the Supply and Transit of Oil from the Persian Gulf

Estimates of the incremental costs to the U.S. defense budget for protecting sources of oil and the routes along which oil is shipped are open to debate, with estimates in the literature ranging from zero to half of the U.S. defense budget. Our estimates indicate that the United States might be able to save between 12 and 15 percent of the fiscal year (FY) 2008 U.S. defense budget if all concerns for securing oil from the Persian Gulf should disappear. However, the size of the residual force would be dictated by remaining U.S. interests in the region.

Policies to Mitigate Threats and Costs to U.S. National Security from Imported Oil

In light of these findings, the United States would benefit from policies that diminish the sensitivity of the U.S. economy to an abrupt decline in the supply of oil. The United States would also benefit from policies that would push down the world market price of oil by curbing demand or increasing competitive supplies of oil, domestic and foreign, and alternative fuels. U.S. terms of trade would improve, to the benefit of U.S. consumers; rogue oil exporters would have fewer funds at their disposal; and oil exporters that support Hamas and Hizballah would have less money to give these organizations. The United States might also benefit from more cost-sharing with allies and other nations to protect Persian Gulf oil supplies and transport routes.

Policies that attempt to curtail the likelihood of an oil embargo against the United States or to reduce oil prices to curb terrorist financing are unnecessary or unlikely to

Table S.1
Potential Links Between Imported Oil and U.S. National Security

Potential Link	Risk or Cost
Large disruption in global supplies of oil	Major
Increases in payments by U.S. consumers due to reductions in supply by oil exporters	Major
Use of energy exports to coerce or influence other countries in ways detrimental to U.S. interests	Minimal
Competition for oil supplies among consuming nations	Minimal
Increased incomes for "rogue" oil exporters	Moderate
Oil-export revenues that finance small terrorist groups	Minimal
Oil-export revenues that finance Harakat al-Muqawamat al-Islamiyyah (Hamas), Hizballah	Moderate
U.S. budgetary costs of protecting oil from the Persian Gulf	Moderate

be effective. Oil embargoes have been an ineffective tool for advancing foreign policy goals. Terrorist attacks cost so little to perpetrate that attempting to curtail terrorist financing through measures affecting the oil market will not be effective.

Policies to Cushion Disruptions in the Supply of Oil

Option: Support well-functioning oil markets. Well-functioning domestic and international petroleum markets are a primary means by which the economic costs of disruptions in the supply of oil can be minimized. Energy prices that are free to adjust to changes in supply and demand, undistorted by subsidies or price controls, offer the most effective mechanism for allocating petroleum in a time of scarcity. Hence, the U.S. government should refrain from imposing price controls or rationing during times of severe disruptions in supply.

Option: Draw on the Strategic Petroleum Reserve (SPR). Releasing oil from the SPR, coupled with coordinated releases from stockpiles in other oil-consuming countries, could completely or almost completely offset the effects of most modest disruptions to U.S. oil supplies. However, U.S. policy for use of the SPR is ambiguous, reducing its efficacy. Currently, the SPR can be used only after a presidential declaration of a "national emergency," which is left undefined. The absence of a publicly stated policy on when the SPR will be used has the potential to trigger panic hoarding if market participants fear a major supply disruption, bringing on the very conditions that SPR use is supposed to ameliorate. By issuing a public statement that the SPR will be used in the event of a major disruption in supply, the market would be better informed and likely act more temperately if such an event came to pass.

Policies to Expand Domestic Sources of Supply

Any measures that increase the long-term global supply of refined oil products or close substitutes will reduce the market power of oil-exporting countries, thereby lowering the world market price of oil. Lower oil prices not only benefit U.S. consumers; they also reduce incomes for rogue oil exporters and potentially contributions to organizations like Hamas and Hizballah, thereby enhancing U.S. national security.

Option: Open access to environmentally sensitive and other restricted areas. Increases in the price of oil have spurred calls to relax or eliminate restrictions on oil exploration and drilling in the Arctic National Wildlife Refuge (ANWR) in Alaska and on the Outer Continental Shelf (OCS) off both the east and west coasts of the United States. A recent study released by the Energy Information Administration (EIA, 2008a) suggests that, if ANWR were to be opened up for oil and natural-gas drilling, it would take approximately 10 years for oil production to begin. At their peak, expanded access to ANWR and offshore coastal reserves might add supply equal to between 4 and 11 percent of baseline forecasts of U.S. demand, reducing future U.S. imports by the same amount.

Option: Increase supplies of unconventional fossil fuels. Unconventional fossil fuels can be produced from coal, oil shale, oil sands, and stranded natural gas. With the exception of Canadian oil sands, production of unconventional fuel substitutes for oil is currently small. However, output from Canadian oil sands and U.S. coal-to-liquid (CTL) plants could be enough to supplant 15 percent or more of baseline domestic U.S. demand for oil. A potential constraint to achieving large production increases is the availability of water and environmental effects. Expansion of CTL will also depend on the costs of controlling—or penalties for releasing—carbon dioxide. CTL is about twice as carbon dioxide–intensive as conventional oil when one factors in all the carbon dioxide emitted, from when it is pumped out of the ground to when it is consumed by a car or truck—that is, on a well-to-wheels basis.

Option: Increase supplies of renewable fuels (biofuels). At present, ethanol produced from corn and blended into gasoline is the most widely used renewable liquid fuel in the United States and is likely to continue to be so. Using corn for ethanol is economically inefficient and has harmed U.S. national security. Diverting corn from food to ethanol production has pushed up world market prices for grains and other foods, which, in 2008, resulted in riots in a number of developing countries. In addition, the net energy benefit of corn-based ethanol is low because so much energy is used to fertilize, harvest, and transport corn. Substantial additional growth in the output of ethanol will have to come from woody, noncrop cellulosic feedstocks (e.g., brush or stubble left after harvest) for which major technological breakthroughs are needed.

Policies to Reduce Domestic Consumption of Oil

Like increases in supply, reductions in domestic petroleum demand put downward pressure on oil prices. However, whereas increases in supply result in an increase in the quantity of oil consumed, measures to increase energy efficiency reduce demand for oil. Greater efficiency reduces the United States' vulnerability to price shocks because generating the same economic output requires less oil. However, like supply-side measures, policies that discourage consumption take a long time to have a substantial effect on demand because improving energy efficiency often requires large investments.

Option: Impose excise taxes on oil. Raising fuel taxes is the most direct way to curb U.S. consumption of oil. Less consumption would put downward pressure on world market oil prices, reducing some of the national security costs linked to U.S. consumption of imported oil. Although prices for U.S. consumers would be higher, net import payments for the country as a whole would be lower, because imports would be reduced.

Even though excise taxes are more effective than other policy measures to encourage more efficient use of oil, fuel taxes have been politically unpopular in the United States, even though the United States has the lowest fuel taxes of any industrial country. How tax revenues from increased fuel taxes would be used would affect their overall economic impact and political opposition as well. For example, a per capita refund of revenues from fuel taxes through the U.S. income-tax system or identifiable improvements in transportation infrastructure would defuse some opposition.

Option: Raise Corporate Average Fuel Economy (CAFE) standards. The economic effects of fuel-economy standards are subject to debate. Proponents argue that these policies overcome market barriers facing consumers who prefer better fuel economy: Fuel-economy standards induce manufacturers to produce vehicles that are in the long-term economic interest of consumers. Other economists have focused on the costs to manufacturers of producing and selling vehicles when consumers may prefer less fuel-efficient vehicles. One study found that increasing gasoline taxes would reduce gasoline consumption for about one-sixth the welfare cost of a corresponding increment to the CAFE standard.

Policies to Reduce U.S. Expenditures to Defend Oil Supplies from the Persian Gulf

The United States could encourage allies to share the burden of patrolling sea-lanes and ensuring that oil-producing nations are secure.

Effective Energy Policies and U.S. National Security

Importing oil imposes costs affecting the national security of the United States. Of the measures we consider in this study, the adoption of the following energy policies by the U.S. government would most effectively reduce these costs:

- Support well-functioning oil markets and refrain from imposing price controls or rationing during times of severe disruptions in supply.
- Initiate a high-level review of prohibitions on exploring and developing new oil fields in restricted areas in order to provide policymakers and stakeholders with up-to-date and unbiased information on both economic benefits and environmental risks from relaxing those restrictions.
- Ensure that licensing and permitting procedures and environmental standards for developing and producing oil and oil substitutes are clear, efficient, balanced in addressing both costs and benefits, and transparent.
- Impose an excise tax on all oil, not just imported oil, to increase fuel economy and soften growth in demand for oil.
- Provide more U.S. government funding for research on improving the efficiency with which the U.S. economy uses oil and competing forms of energy.

Acknowledgments

One of the pleasures of writing this monograph has been our interaction with the many people who have helped us with comments and advice. We would like to express our thanks and appreciation for their assistance. Early in the project, former secretary of defense Harold Brown and Gen. (ret.) John J. Sheehan (U.S. Marine Corps) provided extensive advice on individual chapters and the overall thrust of the report. Gregory F. Treverton of RAND and Hillard Huntington of Stanford University provided two excellent formal reviews. Karen Elliott House provided very useful feedback. We would also like to thank Duncan Long for assistance with data collection and Joya Laha and Christine Galione for support with research, data, and putting this document together. Finally, Frederick C. Smith of the Institute for 21st Century Energy has played an extraordinary role in bringing this project to fruition. He provided thoughtful comments, gently prodded the authors to finish the project in a timely manner, and graciously managed the many different views concerning the goals and subject matter of the report.

Abbreviations

AEO	Annual Energy Outlook
Agip	Azienda Generale Italiana Petroli
ALBA	Bolivarian Alternative for the Americas
ANWR	Arctic National Wildlife Refuge
AOC	Arabian Oil Company
bpd	barrels per day
CAFE	Corporate Average Fuel Economy
CIS	Commonwealth of Independent States
CNOOC	China National Offshore Oil Corporation
CNPC	China National Petroleum Corporation
COCOM	combatant command
CRS	Congressional Research Service
CTL	coal to liquid
CTV	Confederación de Trabajadores de Venezuela
DoD	U.S. Department of Defense
EEC	European Economic Community
EEED	Environment, Energy, and Economic Development Program
EIA	Energy Information Administration
EPA	U.S. Environmental Protection Agency
EU	European Union

FARC	Fuerzas Armadas Revolucionarias de Colombia–Ejército del Pueblo
FDI	foreign direct investment
FY	fiscal year
GAO	U.S. General Accounting Office until 2004; U.S. Government Accountability Office since then
GCC	Gulf Cooperation Council (formally, Cooperation Council for the Arab States of the Gulf)
GDP	gross domestic product
GNPOC	Greater Nile Petroleum Operating Company
GTL	gas to liquid
Hamas	Harakat al-Muqāwamat al-Islāmiyyah
IEA	International Energy Agency
IRGC	Islamic Revolutionary Guard Corps
ISDP	International Security and Defense Policy Center
ISE	RAND Infrastructure, Safety, and Environment
JAM	Jaish al Mahdi
JAPEX	Japan Petroleum Exploration Corporation
JNOC	Japan National Oil Corporation
JOGMEC	Japan Oil, Gas, and Metals National Corporation
mbd	million barrels per day
METI	Ministry of Economy, Trade, and Industry
mpg	miles per gallon
MTW	major theater war
NATO	North Atlantic Treaty Organization
NIOC	National Iranian Oil Company
NOC	national oil company
NPC	National Petroleum Council

NSD	National Security Directive
NSRD	RAND National Security Research Division
O&M	operations and maintenance
OCS	Outer Continental Shelf
OECD	Organisation for Economic Co-Operation and Development
OIF	Operation Iraqi Freedom
OPEC	Organization of the Petroleum Exporting Countries
PDVSA	Petróleos de Venezuela S.A.
PNA	Palestinian National Authority
RDF	Rapid Deployment Force
RDTE	research, development, test, and evaluation
RFS	Renewable Fuels Standard
SACEUR	Supreme Allied Commander Europe
Sinopec	China Petroleum and Chemical Corporation
SODECO	Sakhalin Oil and Gas Development Company
SPR	Strategic Petroleum Reserve
SWA	southwest Asia
TeleSUR	La Nueva Televisión del Sur
UN	United Nations
USAF	U.S. Air Force
USAFRICOM	U.S. Africa Command
USCENTCOM	U.S. Central Command
USEUCOM	U.S. European Command
USGS	U.S. Geological Survey
USN	U.S. Navy
USNORTHCOM	U.S. Northern Command
USPACOM	U.S. Pacific Command

USSOCOM	U.S. Special Operations Command
USSOUTHCOM	U.S. Southern Command
USSTRATCOM	U.S. Strategic Command
WMD	weapons of mass destruction

CHAPTER ONE

Introduction

Purpose

In his 2007 State of the Union address, President George W. Bush stated that U.S. reliance on foreign oil has rendered the nation's interests "vulnerable to hostile regimes, and to terrorists who could cause huge disruptions of oil shipments, and raise the price of oil, and do great harm to our economy." Concerns about the geopolitical and national security consequences of U.S. dependence on foreign sources of oil have triggered arguments for adopting policies to reduce oil imports. Many members of Congress have advocated "energy independence" for the United States so as to reduce potential threats from imported oil to U.S. national security.

Policies focused on reducing U.S. imports of oil ought to rest on a solid conceptual and empirical understanding of both the seriousness of the threats to U.S. national security and the degree to which reductions in U.S. oil imports might mitigate those threats. The purpose of this monograph is to critically evaluate links commonly suggested by political leaders and commentators between imported oil and U.S. national security and to assess the costs and benefits of potential policies to alleviate challenges to national security linked to imported oil. We focus on the following potential links between imported oil and U.S. national security:

- economic
  - the potential for an abrupt fall in supply and the concomitant surge in the world market price of oil to disrupt U.S. economic activity to the point of precipitating an economic recession
  - damage to critical nodes in the U.S. supply chain for refined oil products that could induce short-run local shortages or, if extensive enough, national shortfalls in refined oil products, resulting in a reduction in U.S. economic output
  - large increases in payments by U.S. consumers of oil due to shifts in oil prices because of deliberate reductions in supply by major exporters
- political
  - the potential of major oil exporters to manipulate exports to influence other countries in ways inimical to U.S. interests

- the potential for competition for oil supplies to exacerbate international tensions or disrupt international oil markets
- the effect of higher revenues from oil exports on the ability of "rogue" oil exporters, such as Venezuela and Iran, to thwart U.S. policy goals
- the potential role of oil-export revenues in supporting terrorist groups
- military: the incremental costs to the U.S. defense budget of forces fielded to protect the supply and transit of oil.

The Monograph

We take as our point of departure the conventional wisdom that these potential threats and costs pose significant threats that can be ameliorated by U.S. policies that reduce imports of oil. We draw on both written materials and selected interviews with key experts in the field, as well as our own analysis, to identify what can be said about the nature of the threats (pro and con) and what key elements remain uncertain. We incorporate this analysis into four substantive chapters addressing potential threats and costs, outlined in the remainder of this chapter.

Potential Economic Costs to the United States Posed by Imported Oil

Chapter Two addresses two questions: (1) What is the potential economic effect on the U.S. economy of an abrupt drop in the supply of oil? And (2) to what extent would world market oil prices fall (and the terms of trade improve for the United States) following a decline in U.S. demand for oil? The chapter approaches these questions by first describing the role of oil in the U.S. economy and the linkages between the U.S. market and the global market for oil. It then assesses the major drivers of global demand and supply of oil. It lays out the implications of the fungibility of oil for security of supply and examines the current and past resilience of different parts of the supply chain to shocks. It also compares U.S. reliance on imports to satisfy domestic demand for oil with similar shares for other strategically important commodities. It evaluates the likelihood of a major disruption in international supplies of oil and estimates the potential economic cost of such a disruption to the United States in terms of a decline in U.S. gross domestic product (GDP). It also assesses the sensitivity of world market oil prices to reductions in demand or increases in supply.

Oil as a Foreign Policy Tool

Chapter Three assesses the possibility that oil exporters would use embargoes or subsidized sales of oil as political weapons against the United States or its allies. So as to provide a broader array of examples of embargoes being used as a political weapon, this chapter also looks at instances in which cutoffs of natural-gas exports have been used for political purposes, despite the substantial differences between markets for oil and natural gas. The chapter also reviews the costs and political consequences of attempts by consuming nations to lock up sources of supply.

Oil-Export Revenues, "Rogue States," and Terrorism Financing

Chapter Four evaluates the effects of higher oil prices on the ability of countries to pursue policies antithetical to U.S. interests by examining two cases: Iran and Venezuela. In each case, the chapter looks at the role of oil in government revenues and the ability of these governments to channel revenues to pursue foreign policies aimed at thwarting U.S. interests. The chapter also examines the durability of these policies. The potential role of oil-export revenues in generating funds for terrorist cells, such as al Qaeda, and in financing larger violent movements, such as Harakat al-Muqāwamat al-Islāmiyyah (Hamas) and Hizballah, is assessed.

Incremental Costs of Fielding U.S. Forces to Protect Oil Supplies and Supply Routes from the Persian Gulf

Chapter Five provides estimates of the additional costs to the U.S. defense budget of fielding forces to protect supplies and supply routes for oil from the Persian Gulf. The chapter first discusses different approaches to identify U.S. forces engaged in ensuring that oil supplies and oil supply routes are secure. It then estimates the costs of that portion of these forces that the United States might forgo in the event that the mission to protect oil from the Persian Gulf were to be entirely abandoned. Because of the ambiguities inherent in assigning U.S. military forces to a single mission, the chapter provides a range of cost estimates.

Benefits to National Security and Costs of Policies from Diversifying Sources of Supply and for Reducing U.S. Imports of Oil

Chapter Six critically evaluates various policy proposals for expanding supply and reducing U.S. imports of oil or mitigating the consequences of oil-supply disruptions. It first assesses proposed policies in terms of their likely effects on domestic energy production, domestic oil consumption, and world market oil prices. It also investigates potential impacts, such as increased carbon dioxide emissions from operating synthetic-fuel plants. It then assesses the likely effects of these policies for U.S. national security and the associated costs of these policies. It concludes with a discussion of some of the potentially more-promising and more-effective policy options to secure U.S. oil supplies and mitigate the negative effects for U.S. national security of imported oil.

CHAPTER TWO

Oil Markets and U.S. National Security

Potential Economic Threats to U.S. National Security from Importing Oil

We examine three channels through which changes in the supply of imported oil might affect U.S. national security:

- the potential for an abrupt fall in supply and the concomitant surge in the world market price of oil to disrupt U.S. economic activity to the point of precipitating an economic recession
- damage to critical nodes in the U.S. supply chain for refined oil products that could induce short-run local shortages or, if extensive enough, national shortfalls in refined oil products, resulting in a reduction in U.S. economic output
- large increases in payments by U.S. consumers of oil due to shifts in oil prices because of deliberate reductions in supply by major exporters.

To evaluate the likelihood and potential costs of these threats, we first describe the role of oil, imported and domestic, in the U.S. economy and the linkages between the U.S. and global oil markets.

The Role of Oil in the U.S. Economy

Demand

The U.S. economy moves on gasoline, diesel, jet, and bunker fuels.[1] Roughly 70 percent of the oil consumed in the United States is used for transportation (Figure 2.1). Although other fuels play important roles in electric-power generation, industry, household heating, and chemicals, refined oil products remain ideally suited for transportation. Because they are liquids, refined oil products can be transported and handled

---

[1] Bunker fuel is the fuel used to power oceangoing ships. It is also referred to as *heavy fuel oil* or *residual fuel oil*.

Figure 2.1
U.S. Demand for Petroleum, by Sector

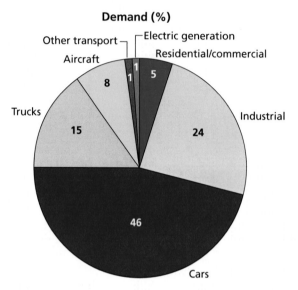

SOURCE: EIA (2008j).
NOTE: Breakdown for cars and trucks was imputed from
figures for motor vehicle gasoline and diesel consumption.
RAND MG838-2.1

easily. They also combust easily and pack large amounts of energy into a relatively small volume.

Demand for transportation drives demand for oil. Although industry consumes about a quarter of U.S. oil, much of this consists of residual fuel oil, petroleum coke, and asphalt, byproducts that remain after refineries have converted whatever they can into transportation fuels. These byproducts typically compete with coal and other fuels for industrial purposes and are priced accordingly.

Imports

On a net basis, the United States imported 12.0 million barrels per day (mbd) of oil in 2007, 58 percent of U.S. consumption (Figure 2.2). In 1973, imports accounted for only 35 percent of U.S. oil consumption; they hit a post-1973 low of 27 percent in 1985. If reference-case Energy Information Administration (EIA) projections of U.S. demand and output materialize over the course of the next two decades, the share of imported oil in U.S. consumption will fall to 53 percent in 2020, but then climb to 56 percent in 2030, slightly less than in 2007.

The largest supplier of imported oil to the United States is Canada, followed by Saudi Arabia and Mexico. Because it is cheaper to transport oil to the United States from Canada, Mexico, western Africa, and Venezuela than from the Persian Gulf, U.S. importers often turn to these sources before turning to the Persian Gulf states.

Figure 2.2
U.S. Consumption and Net Imports of Petroleum and Other Liquid Hydrocarbons

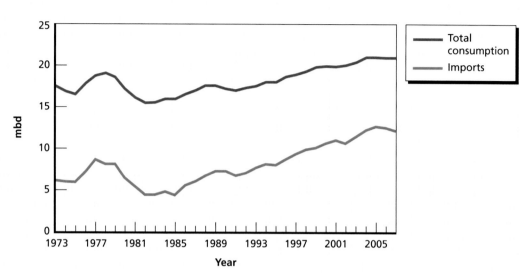

SOURCES: EIA (2008g, 2008h).
RAND *MG838-2.2*

The share of oil imported into the United States is not unusual when compared with other strategic commodities. The share of imported uranium in total U.S. consumption runs 80 percent; zinc, 63 percent; nickel, 60 percent; and aluminum, 44 percent (USGS, 2007, p. 6). However, the value of oil imports dwarfs those of these commodities. Imports of oil and refined oil products totaled $333 billion in 2007, accounting for 16.5 percent of total U.S. imports. Of this total, $253 billion consisted of imports of just crude oil. In contrast, imports of aluminum were only $4.4 billion, and imports of uranium, zinc, and nickel were less than $1 billion combined.

Imports as a share of oil consumption have been rising for decades despite a number of legislative initiatives to increase U.S. production and curb growth in demand. Both Republicans and Democrats have stated that the United States should pursue "energy independence." Energy independence was the primary rationale advanced in support of major provisions of the Energy Independence and Security Act of 2007 (Pub. L. No. 110-140). Definitions of *energy independence* vary from eliminating all use of imported oil to halting the rise in the share of imports in U.S. consumption.

Because the United States still produces substantial amounts of oil, eliminating oil imports might appear, superficially, feasible. However, the gap between U.S. production and consumption is so large that eliminating it would entail extraordinarily costly changes to patterns of consumption and production of fuels. Even if total U.S. oil consumption were to drop substantially, imports would still comprise a large share of the total, in the absence of policies to explicitly restrict imports. Most other net importers of oil do not consider eliminating oil imports to be a policy option. In

Europe, the United Kingdom and Norway produce sizable amounts of oil from the North Sea, but continental European Union (EU) member states produce very little. Japan and South Korea rely on imports for virtually all their oil. For these states, imports are a fact of life.

As discussed in detail in Chapter Six, achieving the more modest policy goal of reducing U.S. imports of oil may be achievable but would still be challenging. *Energy independence*, if defined as eliminating all imports of oil, is not currently an economic option for the United States. This state of affairs is unlikely to change over the course of the next few decades.

Global and U.S. Consumption

The trend in world oil consumption has been up. Despite a dip in the mid-1980s, global consumption rose from 63.1 mbd in 1980 to 85.8 mbd in 2007 (Figure 2.3). U.S. consumption also rose between 1980 and 2007, from 17.1 mbd to 20.7 mbd. However, along with global consumption, U.S. consumption dipped in 2008. The United States remains the largest consumer of oil in the world, accounting for 24 percent of global consumption in 2007. However, the share of the United States and the rest of the developed world—particularly, the EU, Australia, and Japan—has been falling. In 2007, the developed world consumed 47 percent of the global total, down from 58 percent in 1980. In the case of the United States, its share of global consumption has fallen 3 percentage points since 1980, when it was 27 percent.

Figure 2.3
World Oil Consumption

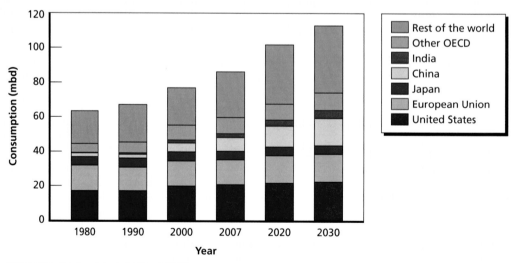

SOURCES: EIA (undated, 2009a, 2008f).
NOTE: OECD = Organisation for Economic Co-Operation and Development.
RAND *MG838-2.3*

Since 1980, consumption in the developed world has grown by only 0.3 percent per year. However, oil consumption by the rest of the world has been growing 2.0 percent per year. Increases in consumption by China have been an important contributor to this trend: China has been increasing its consumption of oil at an average annual rate of 5.5 percent since 1980. It now accounts for 9 percent of global consumption, 1.5 times that of Japan. The rapid increase in oil consumption in China and other developing countries coupled with continued modest growth in consumption in the United States contributed heavily to the sharp rise in oil prices between 2003 and the first part of 2008.

As in the United States, most of the oil consumed in the rest of the world is used for transportation. In the developing world, increases in truck transportation have driven up demand for diesel, and large increases in car ownership have pushed up demand for gasoline as well as diesel. Refined oil products are also used extensively in the developing world to generate electricity, much more so than in the United States or western Europe.

Major energy-forecasting institutions project continued increases in global oil consumption (EIA, 2008d; IEA, 2007; Shell, 2008). According to the EIA, the statistical arm of the U.S. Department of Energy, global oil consumption will rise 32 percent between 2007 and 2030 to 112.5 mbd under their reference-case price scenario (EIA, 2008c). Almost all of the increment (89 percent) will be driven by increased demand from developing countries. Rising economic output and higher incomes in these countries are projected to lead to increased expenditures on automobiles, air travel, and other goods and services that drive consumption of refined oil products.

Increases in the price of oil in 2007 and the first half of 2008 caused some forecasters to scale back their projections of growth in consumption. For example, in 2007, the EIA projected global consumption in 2030 of 118 mbd, 5 percent more than in the 2008 forecast (EIA, 2007c, p. 29). However, even if oil prices return to their levels of the first half of 2008, rising output and incomes will increase global demand for transportation, leading to more consumption of refined oil products until such time as alternative fuels and transportation technologies are substantially more pervasive.

Global Production and Reserves

Global production of oil and liquid hydrocarbons was 84.4 mbd in 2007 (Figure 2.4). Of that amount, the United States produced 8.5 mbd, 10.0 percent of the total. The members of the Organization of the Petroleum Exporting Countries (OPEC) accounted for 38 percent of global production; Saudi Arabia alone supplied 12.1 percent. Output from the former Soviet republics, primarily Russia, Azerbaijan, and Kazakhstan, accounted for 14.9 percent, more than Saudi Arabia. U.S. output has been gradually falling since 1985. In contrast, after sharply cutting production in the 1980s, in 2007,

Figure 2.4
World Oil Production

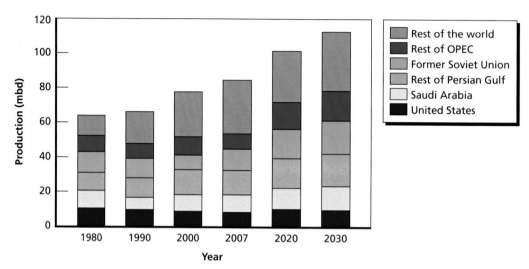

SOURCES: EIA (undated, 2009a, 2008e).
RAND *MG838-2.4*

Saudi Arabia produced as much as it did in 1980. However, during the 2008–2009 recession, it has once again cut output. Russia and the other former Soviet republics have boosted output above past peaks.

The EIA projects that U.S. production will rise through 2020 because of enhanced recovery techniques and more offshore production (EIA, 2007a, p. 95). After 2020, output is projected to decline slightly, but output is still projected to be higher in 2030 than in 2007 (Figure 2.4). Outside the United States, increases in production are projected to come from OPEC, especially the states bordering the Persian Gulf, Russia and other former Soviet republics, and new producers in Africa and Latin America. These suppliers will provide the increase in oil output that will be needed to satisfy the rise in global consumption through 2030.

The National Petroleum Council (NPC) presents estimates from 2006 of projections for oil production for 2030 ranging from a low of 88 mbd from the Association for the Study of Peak Oil and Gas to an average of 107 mbd from the major international oil companies' internal projections to 118 mbd from the EIA's reference-case projections (NPC, 2007, pp. 93, 113). Shell, which published its 2008 oil-output projections in the form of two scenarios, projects output growth through 2020 that closely tracks EIA's 2008 projections. However, for 2030, Shell projects output levels 11 and 17 percent less than EIA's reference case (Shell, 2008, p. 46).[2]

---

[2]   Calculated from figures in exajoules from Shell and converted into mbd using EIA global output numbers for 2000.

Differing assumptions about the extent and effectiveness of government policies to curb oil consumption and emissions of carbon dioxide are one reason for the differences between these projections. Shell chief executive officer Jeroen van der Veer, along with some others, argues that most of the oil that has been easiest to extract will be gone by the end of the next decade (Teslik, 2008). According to van der Veer, new finds will be more difficult and expensive to extract. In contrast, the NPC points out that the cost of new production technologies, such as offshore drilling and enhanced recovery techniques, tends to drop as suppliers and oil producers become more proficient, making difficult fields more economical to develop. In the medium term, even if the world has reached the end of "easy oil," global production is unlikely to decline unless projected demand is much less than currently forecasted. Faced with the right prices, oil companies are likely to meet increased demand for oil even if average extraction costs are higher than they have been in the recent past.

The Persian Gulf is home to more than half of global reserves; members of OPEC own two-thirds (Figure 2.5). Longer term, output will increasingly shift to these countries, especially the Persian Gulf states, where extraction costs tend to be lower.

By definition, *recoverable reserves* are estimated assuming a specific set of technologies and economic conditions. When oil prices rise or technology improves to lower the cost of accessing formerly uneconomic resources, proven reserves rise. For example, many reserve estimates now include Canadian oil sands, which boost Canada's reserves

Figure 2.5
Global Reserves of Oil

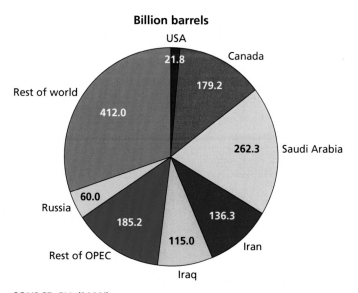

SOURCE: EIA (2008i).
NOTE: Oil includes crude oil and condensate.
RAND *MG838-2.5*

to 181 billion barrels, second only to Saudi Arabia's. If one takes into account other unconventional sources of petroleum, such as shale oil, coal converted to hydrocarbon liquids, and other unconventional forms of fuel that are likely to be tapped over the next 25 years, OPEC's share of global reserves would be appreciably lower.

Prices

International Oil Markets and World Market Oil Prices

Because oil can be transported easily, oil markets are much more globally integrated than those for electric power and natural gas, in which access to sources of supply is constrained by transmission lines or by pipelines and expensive liquefaction and regasification facilities. Most of the world's largest refineries have easy access to seaborne oil directly or by pipeline. Once oil reaches a port for shipment, it can be easily transported to these refineries. Whereas regional supply and demand drive prices for electricity and natural gas, prices of benchmark oils, such as West Texas Intermediate or Brent Crude, are determined by trading on international markets.

These benchmark oils tend to be *sweet:* low in sulfur and other contaminants. They can be easily processed by all major refineries and are therefore widely traded. Heavy and *sour* oils are costlier to refine because they require special equipment and more-extensive processing. The discount is determined by the capacity of those refineries as well as by the price of benchmark crudes.

Historically, crude oil prices were often determined by long-term contracts. In the 1950s and 1960s, the Seven Sisters, the seven largest international oil companies at the time,[3] negotiated long-term, fixed-price contracts with the Middle Eastern states. In the United States, the Railroad Commission of Texas regulated producers in Texas, the state in the United States that produced the most oil, forcing them to adjust oil output to avoid "ruinous" price-cutting competition. Because the majors were vertically integrated—that is, they explored, produced, transported, refined, and sold gasoline and diesel to retail customers—changes in demand and prices in retail markets affected refinery and retail margins, but the majors attempted to insulate oil prices from these shifts by confining price adjustments to the downstream parts of their businesses. Consequently, wellhead prices remained fairly stable during this period, in great part because increased production from new finds kept pace with increased global demand.[4]

---

[3]   Standard Oil of New Jersey (Esso; later, Exxon), Royal Dutch Shell, Anglo-Persian Oil (now BP), Standard Oil of New York (Socony; later, Mobil; now part of ExxonMobil), Standard Oil of California (Socal; later, Chevron), Gulf Oil, and Texaco (now part of Chevron).

[4]   The wellhead price is the price of oil in the field, immediately after it emerges from the ground.

In the 1970s, inflation, the collapse of the Bretton Woods system of fixed exchange rates, shifts in supply and demand, and the 1973–1974 oil embargo resulted in more volatility in prices (Yergin, 1991). One of the first harbingers of increased volatility in oil prices was the Texas Railroad Commission's decision to abandon controls on supply in 1970. OPEC members began to exert more influence over prices and production decisions, nationalizing reserves from the international companies. When the Middle Eastern oil producers embargoed sales to the United States, the Netherlands, and Portugal in 1973 and 1974 and cut production by 25 percent, oil prices soared (Figure 2.6). Prices jumped again after the outbreak of the Iran-Iraq War, peaking in 1980. They then plummeted in the 1980s as new supplies came on stream and demand fell due to higher prices. In real terms, oil prices did not reach 1980 levels again until 2007.

Spot Markets. The way in which oil prices are determined has dramatically changed since the 1960s. Over the course of the past four decades, spot markets in London and New York have developed from small secondary markets, used for making up shortfalls in deliveries or selling excess crude, to become the primary means for determining oil prices. Today's highly liquid markets bring together a network of buyers and sellers who use a wide variety of trading instruments. Contracts run in the tens of millions every day. With the advent of international markets, differences in prices across regions, other than those justified by differences in transportation costs or oil "quality" (e.g., specific gravity, viscosity, sulfur content), cannot persist: Traders arbitrage such differences away. Because of traders' ability to arbitrage, the markets' influence on

Figure 2.6
Price Per Barrel of West Texas Intermediate in Current and Year 2000 Dollars

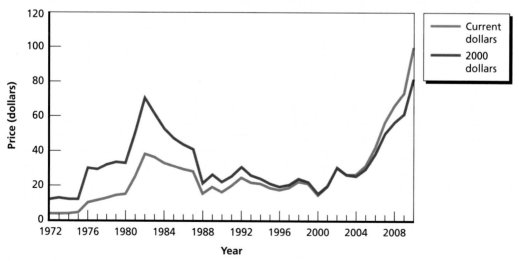

SOURCES: IMF (undated); EIA (2009b); BEA (2008).
RAND *MG838-2.6*

oil prices goes far beyond just the oil traded on the markets. In addition, prices of oil sold under longer-term contracts are now closely tied to prices on the spot markets.

The fungibility of oil has implications for energy security whose importance cannot be overstated: From an economic perspective, where the United States acquires its oil has become irrelevant. Disruptions of supplies or jumps in demand anywhere in the world will be distributed across the world market. Conversely, attempts by foreign suppliers to target supply reductions toward specific importers cannot succeed because oil will be sold on through the markets to the highest bidder, whoever that may be.

Futures Markets. The growth of spot markets for crude oil and refined oil products has set the stage for the other major development in the structure and operation of the international oil markets: the emergence of high-volume futures trading and other forms of derivatives.[5] Futures contracts typically do not result in actual delivery or acceptance of the product. They provide a means for both sellers and purchasers to hedge risks of movements in oil prices that they would find unfavorable. Both spot and futures markets allow price signals to be transmitted quickly across time as well as space: An expectation of future supply constraints will quickly be reflected in both today's futures prices and today's spot prices as inventory holders build stocks.

Futures contracts and other derivatives involve speculators who take the opposite position from hedgers in hopes of profiting from unexpected price movements. In the course of the heated debate over the rise in oil prices in 2008, speculators were accused of bidding up prices above levels justified by market fundamentals, creating a bubble. There is some evidence that, when uncertainty is high and some traders follow the actions of others they believe to be better informed, lemminglike behavior may occur in oil markets, contributing to surges in prices up and down (Weiner, 2002). Notwithstanding the potential for bubbles, the existence of futures markets is beneficial because they create opportunities to hedge without having to bear the cost of owning physical stocks of oil.

Oil Price Volatility

The price of oil, like the prices of gold or corn, tends to be more volatile than prices for manufactured goods and services. Because both consumption and output of oil are slow to respond to changes in price, both the demand for and supply of oil are inelastic. Sharp fluctuations in oil prices are due in great part to the inelasticity of demand for and supply of oil.

Elasticity of Demand. As noted in the preceding section, demand for crude oil is driven by demand for transport fuels. Currently, there are few substitutes for refined oil products in transportation. Households find it difficult to forgo many trips—most

---

[5]   In finance, a derivative is a security whose price is dependent on or derived from one or more underlying assets, such as stocks, bonds, commodities, or currencies. Its value is determined by fluctuations in the underlying asset. Futures contracts, forward contracts, options, and swaps are the most common types of derivatives. Derivatives are generally used to hedge risk but can also be used for speculative purposes.

notably, those to work, to school, and to purchase necessities. They continue to use the cars they already own even if those are not very fuel efficient. In the short run (less than one year), estimates of the elasticity of motor-fuel demand in the United States and other developed countries run about −0.1—that is, an increase of 10 percent in the price of gasoline would trigger a fall of 1 percent in consumption, all other factors held constant (Cooper, 2003; Gately and Huntington, 2002; Greene and Ahmad, 2005; Huntington, 1994, 1991). To take a recent example, during the period of January through May 2008, gasoline prices rose 25.5 percent over the same period in 2007, and U.S. gasoline consumption fell 3 percent while incomes more or less stagnated. Using these numbers yields a U.S. price elasticity of demand for gasoline of −0.12.

Faced with higher fuel prices, over time, consumers change their behavior: They purchase more–fuel-efficient vehicles, take more public transportation, and shorten trips. They may even move or change jobs so as to be closer to work. Consequently, long-term elasticities (10 years or more) are larger than short-run elasticities. A low-end estimate for the longer-term elasticity of demand for motor-vehicle fuels in the United States is −0.3; an upper-end estimate, −0.7 (Bartis, Camm, and Ortiz, 2008, p. 153).

Elasticity of Supply. Estimates of global long-term supply elasticities range from 0.3 to 0.5—that is, a 10-percent increase in long-term prices should result in a 3- to 5-percent increase in global supply (Huntington, 1991; Greene and Ahmad, 2005). Short-term supply elasticities are much lower because production capacity is fixed in the near term. Analysts differ on how OPEC and its members are likely to respond to changes in market conditions. OPEC is widely seen to have the potential to exercise market power rather than just respond to market signals.

The key point for U.S. energy security is that short-term supply and demand elasticities are low both in the United States and in the rest of the world. As a consequence, small perturbations in supply or demand can have major effects on oil prices, both up and down, like those in 2008. These impacts can persist for some time until either the perturbation ceases or the market more fully adjusts.

Supply and Demand Rigidities

Policies adopted by governments in both consuming and producing nations have made both demand for and supply of oil less responsive than they would be absent these policies. Price controls and subsidies on refined oil products insulate consumers from world market price increases. State-owned oil companies have been slower than private companies to respond to higher prices by increasing output and exploration.

Refined Oil Product Price Subsidies. Many countries in the developing world sell refined oil products at controlled or subsidized prices. Although this practice is most closely associated with OPEC member states, a number of oil-importing countries, including some of the poorer African states, sell refined oil products at fixed prices. When world market prices rise, these governments are slow to raise domestic prices, either letting tax revenues fall or providing subsidies from the budget to cover the increased costs of imported refined oil products. Some countries cross-subsidize retail prices of refined oil products from domestic production. For example, in China, the state-owned oil producers have to go to the national government and request payments to compensate for losses they incur from selling gasoline and diesel at controlled retail prices or cover their losses from profits on domestic production. In the first part of 2008, these Chinese companies lost money on every barrel of imported oil they refined and sold domestically. Controlled prices and the associated subsidies dampen the effect of increases in the world market price of oil on consumption. Because consumers in these countries do not face the increased cost of oil, they do not reduce consumption, propping up prices on the world market.

State-Controlled Oil Companies and Production Taxes. The countries that control the bulk of global oil reserves limit ownership and often restrict production to state-owned oil companies. Because of lack of competition and corporate goals driven by political rather than commercial concerns, state-owned crude oil suppliers tend to be less efficient than privately owned oil companies. They employ more workers per unit of output, deplete reserves more rapidly, and invest less in capital equipment, maintenance, and research and development (Eller, Hartley, and Medlock, 2007; Jaffe, 2007). These forms of technical inefficiency result in higher costs per unit of output. State ownership may also lower the elasticity of supply: State-owned companies often have less motivation or ability than private-sector firms to expand output when prices rise.

State-owned oil companies are not a new phenomenon and are not located just in producing countries. Governments own controlling stakes in major oil companies in China, India, and most African and Asian countries. Companies, such as Azienda Generale Italiana Petroli (Agip) in Italy, Total in France, and BP in the United Kingdom, were once controlled by the state. These companies were privatized over the course of the past few decades because governments became convinced that they would be run more efficiently in the private sector. Following Europe's lead, a number of developing countries, some of them producing countries, such as Russia, have privatized parts of their oil sectors over the course of the past two decades.

Governments in oil-producing countries tend to raise marginal tax and royalty rates when market prices rise. Lease conditions also become less attractive. Increases in marginal tax rates may dampen responses to changes in price, as they may make some additional production unprofitable. In Russia, marginal export taxes rise sharply when export prices exceed stipulated levels. Above those prices, the state takes almost all the additional revenues. During 2007 and the first part of 2008, a period of historically high oil prices, Russian production stagnated, in part because of the export-tax regime.

Saudi Arabia and some other OPEC members with excess capacity have attempted to prop up oil prices by reducing output (Gately, 2007). They have also been slower to develop new fields than would private-sector companies faced with the same incentives. These actions reduce the global elasticity of supply.

Oil-Market Disruptions and U.S. National Security

Risks

Most of the world's oil reserves are located within Russia, other former Soviet republics, and the member states of OPEC, especially the countries of the Persian Gulf (Figure 2.5). Global oil supplies from these countries have been frequently disrupted because of instability and conflict. In the Persian Gulf, wars have stopped production in one or more countries in each of the past three decades. Iranian production plummeted in the aftermath of the 1979 revolution. Insurgency, political instability, and strikes have periodically brought oil production to a halt in Nigeria and Venezuela. Few other commodities as important as oil have been so subject to disruption from war and political turmoil.

Supply lines for oil are long, involving pipelines and supertankers that pass through narrow maritime passages, such as the Strait of Hormuz, the Strait of Malacca, and the Bosporus and the Dardanelles. Terminals and refineries represent points of potential vulnerability. While pipelines can be put back into service relatively quickly, large oil-handling and -processing facilities could take many months to be repaired.

Some commentators argue that dependence on unstable and potentially hostile foreign suppliers and potentially vulnerable supply routes increase the risk that supplies of oil to the United States might be curtailed. Some commentators have also been concerned that supply disruptions could lead to shortages of fuel available to the U.S. military.

However, the ability of even the most politically motivated government of a major oil-exporting country to sustain a prolonged reduction in oil output is open to question. Production did stop after the Iranian Revolution but the ayatollahs resumed production as quickly as possible; Iran has tended to produce at maximum capacity. Because oil export revenues account for such a large share of the Iranian government's budget revenues and the incomes of residents, a sustained cutoff in export revenues is not financially feasible. Radical regimes have not been willing to cut oil production because they need the revenue.

Blocking the free flow of oil by blockading straits would entail a major military effort: The straits of Hormuz and Malacca are several miles wide at their narrowest points. Other than the United States, few countries could make a credible threat to blockade these straits. Iranian efforts to disrupt flows of oil through the Persian Gulf during the early 1980s were not very successful.

Note that, despite our discussion of the location of oil reserves, a disruption in the global supply of oil is an economic, not geopolitical, problem. In the event of a reduction in supply, the market as a whole shares the burden of the supply reduction, as prices rise to ration demand. In the event of an abrupt reduction in the global supply of oil, the United States would not face a physical shortage, but U.S. consumers would have to pay the higher market price, which would lead to a fall in consumption. By the same token, as long as the U.S. military is willing to bid high enough, it will have access to fuel.[6]

The Energy Modeling Forum at Stanford University has used a panel of energy-market and geopolitical experts to derive probabilities for the occurrence of at least one major disruption in supply over the course of the next 10 years (Beccue and Huntington, 2005). The forum concluded that the probability of a short disruption (one to six months) entailing a reduction in 10 mbd or more during this period is about 8 percent (Figure 2.7). For a short disruption of 15 mbd or more, the probability falls to about 1 percent. Looking at the graph from the vantage of the length of a reduction in supply, the forum concludes that there is an almost 50-percent probability of a short disruption (one to six months) of 5 mbd or more.[7] The probability falls to about

Figure 2.7
Probabilities of Disruptions of Global Oil Supplies: Magnitude and Duration

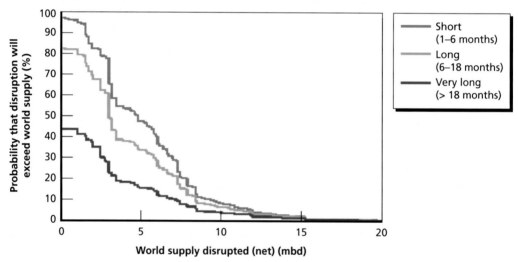

SOURCE: Beccue and Huntington (2005).
RAND MG838-2.7

---

[6] In World War II, the U.S. government chose to ration fuel (along with a number of other products), giving the military precedence. The rationing system restrained the cost of fuel for the government but at the cost of a loss in economic efficiency. Rationing also led to black markets in gasoline and other rationed products.

[7] This decline is analogous to the fall in world oil output of 4.3 mbd between 1979 and 1981 because of the Iranian Revolution and the Iran-Iraq War.

35 percent for a disruption of this size for six to 18 months, and about 15 percent for a disruption of more than 18 months.

Costs to the U.S. Economy of Supply Disruptions

A major national security concern for U.S. policymakers is the potential for an abrupt reduction in the supply of oil and a corresponding large increase in the price to result in a sharp fall in economic output. Such a decline would undermine U.S. national security, for example, by weakening U.S. global economic and political influence and the ability of the United States to pay for U.S. military forces.

The key phrases are "abrupt reduction in supply" and "large increase in the price." Although most recessions in the United States since World War II have followed sudden jumps in the real or inflation-adjusted price of oil, not all price spikes have been followed by recessions. Price increases driven by increases in global demand do not appear to be correlated with recessions (Balke, Brown, and Yücel, 1999). Gradual increases in the real price of oil may not have a measurable effect on growth.

How might an abrupt fall in oil supplies affect the U.S. economy? A sharp fall in global oil supplies would precipitate a rise in world market oil prices. The sharp increase in the price of oil and other forms of energy may make it unprofitable to produce energy-intensive or oil-intensive products. Closing plants that produce these products would contribute to a decline in output. Because the United States is a net importer of oil, an increase in the real (inflation-adjusted) price of oil would also result in a deterioration in the U.S. terms of trade, leaving U.S. consumers and businesses with less buying power than when prices are lower. These reductions in purchasing power or income effects would contribute to less economic activity. Some types of labor may also be priced too high in relation to oil. If real wages do not adjust to reflect this change, a rise in unemployment might accompany the oil price shock. Changes in demand due to changes in the real price of oil would lead to changes in the pattern of demand and adjustment costs as businesses respond to those changes. For example, as consumers shifted toward more–fuel-efficient vehicles in 2007 and 2008, U.S. automakers experienced a sharp decline in demand for full-duty pickup trucks and sport-utility vehicles. U.S. automakers have had to close or retool some of the plants designed to manufacture these vehicles. Scrapping formerly productive capital stock contributes to lower growth.

A major challenge to gauging the potential magnitude of the effect of an abrupt surge in the real price of oil on the U.S economy is that oil price shocks have often been accompanied by other macroeconomic policies that have affected aggregate demand. For example, in both 1973–1974 and 1979–1980, U.S. monetary policy had been inflationary prior to the disruption. The tightening of monetary policy in 1980 affected the severity of the 1980 and 1981–1982 recessions (Bernanke, Gertler, and Watson, 1997; Barsky and Kilian, 2001, 2004).[8]

To translate the previous discussion of potential oil supply shocks into estimates of their potential effect on the U.S. economy, we used the estimates of the elasticity of oil demand and supply presented in the preceding section. Assuming a short-term price elasticity of demand for oil of –0.1 and that the supply of oil is running at 85 mbd, the quantity supplied in 2007, an 8.5 mbd (10-percent) fall in supply would result in a doubling of oil prices so as to match demand with the new lower supply.

Economists have attempted to estimate the potential negative effects of an oil price shock on GDP by estimating the elasticity of GDP with respect to the oil price.[9] A review of the literature on the macroeconomic effects of oil shocks suggests a range for a near-term elasticity of –0.01 to –0.05 (GAO, 2006; Nordhaus, 2007; Beccue and Huntington, 2005; Jones, Leiby, and Paik, 2004; Brown and Yücel, 2002). This range of GDP elasticities implies that a rapid, sustained doubling of oil prices could lead to a decline in GDP of 1 to 5 percent, all other factors being equal. This decline would be superimposed on baseline economic growth: If the economy were growing at 2 percent per year, the shock could translate into growth of 1 percent to a decline in output of up to 3 percent. Larger declines in output (higher elasticities) are associated with sudden oil price shocks; more-gradual increases in the real price of oil yield smaller declines in growth.

The magnitude of the macroeconomic costs due to higher oil prices depends on a number of factors, three of which are particularly important: (1) the magnitude of the price shock, (2) its persistence, and (3), the importance in the overall economy of consumption of oil (all oil, not just imports) and its close substitutes. Larger and longer-lasting shocks are expected to generate larger adverse impacts. Macroeconomic vulnerability is expected to be larger if oil plays a larger economic role, whether oil is domestically produced or imported.

Some recent studies conclude that the higher absolute values in the range of GDP elasticities are influenced by the inclusion in the statistical estimates of the price shocks of the 1970s and early 1980s and that vulnerability today is lower, probably closer to –0.01. The U.S. economy uses less oil and other forms of energy to generate the same amount of output today than in that earlier period; labor and other factor markets are also more flexible.

U.S. economic output is affected by economic activity in the rest of the world. Like the U.S. economy, the global economy is potentially vulnerable to an abrupt decline in oil supplies and a sharp increase in the real price of oil. Using a global macroeconomic model, the International Monetary Fund has simulated the impact of an oil supply shock that leads to a doubling of oil prices (IMF, 2007, pp. 17–18). It finds

---

[8] For counterarguments, see Hamilton and Herrera (2004) and Hamilton (2005).

[9] This relationship is unlikely to be constant across all possible oil price shocks or across time.

that, in this scenario, world GDP falls by 1.4 percent and global inflation increases by 1.5 percent relative to the model's baseline forecast. The model forecasts that the economic slowdown will be most pronounced in emerging Asian economies because those economies are more oil intensive than those of more-developed nations. Oil-exporting nations run a large trade surplus, peaking at around 6 percent of GDP above the baseline, and enjoy a notable economic expansion.

A recent study by Leiby (2007) seeks to calculate the incremental benefit of reducing U.S. oil imports, where the macroeconomic benefit is the expected reduction in vulnerability from an oil supply shock given different assumptions about the size and magnitude of such a shock. Leiby calculates an incremental macroeconomic premium of around $4.70 per barrel of imported oil (in 2004 dollars), with a range of $2.20 to $7.80 per barrel reflecting different uncertainties in his model. This is equivalent to 5 to 17 percent of the base oil price, with an expected value of around 10 percent.[10] This premium on imported oil is the additional cost to the U.S. economy of each barrel of imported oil consumed, stemming from the vulnerability of the U.S. economy to an oil price shock.

Mitigating the Costs of Supply Disruptions

Private-sector use of buffer stocks and policy decisions to use strategic oil reserves, such as the United States' Strategic Petroleum Reserve (SPR) and those kept by other members of the International Energy Agency (IEA), would greatly diminish or even eliminate the actual supply loss felt by the market and thus the price shock. Conversely, if a supply shock were to trigger a wave of panic buying so as to increase private inventories out of fear that conditions could worsen, the impacts of a supply shock would be magnified.

Expanding and diversifying sources of supply through the development of new oil fields or the production of oil substitutes—both unconventional fossil resources and renewable fuels—would make the U.S. economy less vulnerable to supply shocks. Oil sands have become commercially competitive at current oil prices. At prices of $65 per barrel and above, economic opportunities for developing motor-vehicle fuels from unconventional fossil fuels, such as coal or, potentially, oil shale, are significant (Bartis, Camm, and Ortiz, 2008; Bartis, LaTourrette, et al., 2005). Dramatically expanding the production of motor-vehicle fuels from plants requires a technological breakthrough to convert cellulose into alcohol. If a cost-competitive technology becomes

---

[10] These estimates are based on an oil price averaging around $45 per barrel out to 2015 and the disruption risks taken from the Energy Modeling Forum study (Beccue and Huntington, 2005), among other assumptions.

The higher end of these figures reflects a GDP loss elasticity with respect to oil shocks larger in absolute value than the −0.05 figure noted earlier.

Since the premium depends on the size of the potential price shock relative to an equilibrium baseline price, as well as the importance of oil use in the economy, it would be inappropriate to apply these same percentages to a higher baseline price to recalculate a premium applicable to 2009 market conditions.

available, renewables could also contribute substantially to supply (Toman, Griffin, and Lempert, 2008). While the likely alternatives would also have low short-term price elasticities of supply, diversification would imply a lower likelihood of a particular size economic shock as a result in a drop in the supply of oil.[11]

Improvements in the efficiency with which oil is used would also reduce U.S. economic vulnerability to a sudden increase in oil prices. If oil were used more efficiently, a price shock would have a smaller effect on the economy because the role of energy in the economy would be smaller. U.S. economic vulnerabilities are also reduced if demand responds more quickly to changes in oil prices. For example, if consumers could switch easily from commuting by car to mass transit or telecommuting when fuel costs rise, the effects on the U.S. economy of an increase in prices would be smaller.

The Resilience of the Supply Chain

Crude oil traverses a long road between well, refinery, and filling station. The supply chain consists of wells, collection and cleaning facilities near the field, pipelines, terminals, tankers, refineries, product pipelines, product terminals, tanker trucks, and filling stations. A disruption in this supply chain can have adverse effects on the entire refined oil product market. For example, following Hurricane Katrina, gasoline and diesel prices in the United States rose sharply as crude oil from offshore rigs could not be landed, refinery operations near the Gulf of Mexico stopped, and some product pipelines could not operate because of the lack of electric power.

The U.S. and global supply chains have been resilient. Companies have moved quickly to restore facilities that have been damaged. Losses of refining capacity (e.g., during lengthy repairs after a major fire) have required increased product imports or increases in capacity utilization at other refineries. However, the resulting scarcity can be widely shared throughout the market. Similarly, a pipeline breakdown reduces crude or product deliveries, but products can be delivered by barge, train, or truck as well as by pipeline.

One factor that has limited the resilience of the supply chain is market segmentation due either to constraints on refineries' use of different oil grades or to air quality or other regulations and mandates that require specialized product blends in different market areas. These blends are referred to as *boutique fuels* because they are formulated for specific metropolitan markets.

Both of these factors reduce the fungibility of the domestic petroleum market and may lead to more–regionally localized price disturbances. The first has not been

---

[11] Note that, as long as refined oil products remain the dominant form of transportation fuel, the prices of substitutes will be driven by the price of oil. Consequently, even if production of alternative fuels expands, a sudden drop in the supply of oil would have a similar short-run impact on the U.S. economy as it currently does, as the rise in refined oil product prices would cause prices of close substitutes to rise as well.

a major problem in the United States. U.S. refineries tend to be sophisticated and can handle a variety of crudes.

Ensuring supplies of boutique fuels has sometimes been more challenging. Under the Clean Air Act (Pub. L. No. 88-206), state environmental authorities may seek to meet air quality standards by stipulating that locally sold gasoline be formulated so as to produce fewer emissions. As a consequence, gasoline blends differ in urban areas. These differences have sometimes generated localized price shocks: In the summer of 2002, when a refinery was out of commission in the Chicago area, gasoline prices were much higher in Chicago than in Milwaukee, a nearby metropolitan area where air quality standards mandated a different, more–readily available formulation for gasoline.

U.S. Terms of Trade, Oil Prices, and National Security

In 2008, the United States imported 4.9 billion barrels of crude oil and refined oil product equivalent at an average landed price of $94.63. In 2007, the United States paid $67.97 per barrel of imported crude. If prices had remained the same, U.S. consumers would have spent $131 billion less than they did. These increased expenditures are a consequence of changes in what economists call the *terms of trade*, the volume of exports needed to purchase a given volume of imports. Because the United States is a net importer of oil, when oil prices fall, as they did in the second half of 2008, the United States benefits from an improvement in its terms of trade as consumers pay less for oil. When oil prices rise, as they did in 2007 and the first half of 2008, the terms of trade deteriorate because consumers pay more for the same quantity of oil.

Oil is a market in which lower-cost producers have benefited from substantial rents, the difference between costs, including a normal rate of return, and price. Many of the countries bordering the Persian Gulf are among low-cost producers. Most of these countries (Iran, Iraq, Kuwait, Qatar, Saudi Arabia, and the United Arab Emirates) have banded together in OPEC for the purpose of influencing the world market price of oil by adjusting output. The United States, for its part, as the largest consumer of oil in the world, has the ability to put downward pressure on the price of oil if it can reduce consumption.

As noted earlier, global oil demand and supply are inelastic. Reductions in the output of oil by these suppliers can have a substantial impact on price and, by extension, U.S. payments for imported oil (see Bartis, Camm, and Ortiz, 2008). When OPEC has successfully reduced supply and pushed up prices, the U.S. economy has suffered a fall in its terms of trade and U.S. consumers have suffered a loss in purchasing power. These higher prices result in increased expenditures on oil, expenditures that U.S. consumers could have used to purchase other goods or that U.S. businesses could have used to invest in factories and equipment. Alternatively, a fall in demand or increase in non-OPEC production pushes down prices. Lower world market oil prices benefit oil consumers and, by releasing funds for other purposes, provides potential benefits to U.S. national security by making defense dollars go further or boosting U.S. economic activity. For example, the decline in world market oil prices in the second half of 2008 has dramatically reduced U.S. Department of Defense (DoD) expenditures on diesel and jet fuel, easing pressures on the defense budget.

CHAPTER THREE

Oil as a Foreign Policy Instrument

Because of the importance of refined oil products for transportation, politically motivated cutoffs in supplies of oil are a potential national security concern for oil-importing states. In a number of instances, leaders of oil-exporting countries have attempted to exploit this perceived vulnerability, threatening and, in some cases, imposing embargoes on the export of oil. U.S. policymakers have expressed concern about the ability of exporters to buy support for their foreign policies by providing importing countries with free or subsidized fuels. They have also expressed concern that the United States may be prevented from purchasing oil because other consuming nations have locked up oil supplies through investments in producing nations. This chapter critically examines these concerns.

Oil Embargoes and Cutoffs

A number of countries have embargoed the export of oil to countries with which they are at odds. South Africa faced an almost universal official embargo on oil, although it never had problems arranging oil imports through third parties. Most Arab states continue to embargo exports of oil to Israel. Consuming nations have embargoed oil imports: The United States, for example, refuses to purchase oil from Iran. These embargoes have been adopted to pressure foreign governments to change policies, such as apartheid in South Africa, or, in some instances, to precipitate a change in government. In this section, we review the successes and failures of oil embargoes to influence foreign policy decisions by consuming nations. To provide additional insights on the effectiveness of export embargoes, we also examine the political consequences of Russian cutoffs of natural-gas exports, although there are some marked differences between oil and natural-gas markets.

Oil Export Embargoes Prior to and During World War II

Oil was already a vital economic input for industrialized economies prior to World War II. In recognition of oil's importance to the Italian economy, the League of Nations imposed an embargo on sales of oil and refined oil products to Italy in an attempt to force it to halt its invasion of Abyssinia (Ethiopia). Germany and other countries friendly to Italy refused to uphold the embargo: They subverted it by reselling oil and refined oil products to Italy. The League of Nations' sanctions failed.

Concerns about security of supply played important roles in both Japan's and Germany's choices of military objectives in World War II. Japan depended on imports from the United States and the Dutch East Indies for most of its oil. The United States imposed an oil export embargo against Japan in 1941 in an effort to induce Japan to withdraw its troops from China (see, e.g., Levy, 1982, pp. 24–35). Japan's invasion of Southeast Asia and the simultaneous attack on Pearl Harbor were driven, in part, by the Japanese government's desire to ensure supplies of oil by seizing the East Indian fields and crippling the United States' ability to come to their defense.

The German drive through Ukraine to capture the Russian oil fields in the Caspian was driven by the goal of securing oil supplies and to cut off Soviet access. Japanese and German concerns about supplies of refined oil products were not unfounded. Shortages of diesel, gasoline, and aviation gasoline hampered the ability of both countries to prosecute the war.

The 1956 Saudi Arabian Oil Embargo Against France and the United Kingdom

In 1956, a British-French-Israeli military operation wrested control of the Suez Canal from the Egyptians after the Egyptian government nationalized the canal, which had been owned by British and French companies. Saudi Arabia responded by embargoing oil exports to France and Great Britain. (Saudi Arabia already had a policy of refusing to sell oil to Israel.) Because Great Britain and France were able to substitute supplies from other producers, neither experienced supply difficulties. However, when the United States threatened to refuse to make up for the shortfall in oil supplies by increasing its exports to these countries, concerns about the supply of oil rose in both countries.

The three countries did withdraw their forces from Egypt. However, the Saudi oil export embargo is usually regarded as having played a minor role in this decision. Political pressure from the United States and France's and the United Kingdom's European neighbors, as well as domestic political divisions over the wisdom of the operation, coupled with a U.S. threat to withhold financial support in the face of a looming devaluation of the British pound, were of more importance in prompting the decision to withdraw.

The 1967 Oil Embargo Against the United States, the United Kingdom, and Germany

On June 6, 1967, the day after the outbreak of the Six-Day War, Arab countries embargoed oil exports to those Western countries that were providing political or military support to Israel. The main targets of the embargo were the United States, Great Britain, and, subsequently, the Federal Republic of Germany. None of the countries experienced supply shortfalls. World market oil prices did not rise appreciably, in large part because the Arab exporters did not accompany the embargo with a cut in production. Countries that were not targets of the embargo increased imports from the Arab producers, while the three targeted countries were able to substitute supplies from other sources. The embargo lasted almost three months, ending on September 1, 1967, with the Khartoum Resolution.

While the embargo did not inflict substantial economic harm on the three countries, it did set a precedent for collective action on the part of Arab oil producers, which had successfully coordinated a collective embargo on direct sales of petroleum against three major consuming countries. Although the governments of the United States and other Western nations became more aware of the economic vulnerabilities associated with depending on oil imports from Arab producing countries, the 1967 embargo did not trigger major policy responses on their part.

The 1973–1974 Oil Embargo Against the United States, the Netherlands, Portugal, and South Africa

In October 1973, the Arab member states of OPEC imposed an oil embargo against the United States, the Netherlands, and Portugal for supporting Israel during the Yom Kippur War. South Africa was also included in the embargo. OPEC member states also announced a 25-percent cut in overall production. They also raised oil prices: By early 1974, prices had risen to $11.65 per barrel, up from $3 per barrel in 1972. Both the embargo and the production cuts were lifted in March 1974.

The principal goals of the embargo were to compel Israel to withdraw from the territories it had occupied following the 1967 war and to weaken Western support for Israel in its ongoing conflict with Syria and Egypt. It was also intended to force the United States to reduce its political, economic, and military support for Israel. Capitalizing on Europe's dependence on imports of oil from Arab countries, it was designed to enlist European countries in support of these goals—in particular, to induce the Europeans to pressure the United States to change U.S. policies concerning the Middle East.

The United States did start negotiations with Arab oil producers and began a process of shuttle diplomacy among Egypt, Syria, and Israel. This shuttle diplomacy eventually resulted in a January 1974 agreement for Israel to pull back from newly occupied areas of Egypt. In May 1974, a similar agreement was reached on the return of Syrian territory occupied during the Yom Kippur War. These talks also laid the groundwork for the Camp David Accords. In March 1974, the embargo against the United States and its allies was lifted, after the Arab members of OPEC stated that substantial progress in Arab-Israeli disengagement had occurred.

Although the embargo encouraged greater U.S. diplomatic efforts in the Middle East, it did not attain its other goals. It did not lead to a split between the Europeans and the United States over Middle Eastern policies. It did not substantially alter U.S.

policy positions concerning Israel or the Palestinian conflict. When the United States showed no signs of changing its overall policy stance, Saudi Arabia pushed for a change in the oil cartel's policy toward the West to avoid further souring U.S.-Saudi relations. Saudi Arabia took the lead in drafting the formal statement by Arab members of OPEC announcing the end of the embargo.

While the production cuts that accompanied the 1973–1974 Arab oil embargo triggered a sharp rise in global oil prices, observers dispute the actual role of OPEC's pricing policy and the accompanying production cuts on longer-term prices. Some argue that the embargo determined the timing of the price adjustment but not the size. Rising global demand, the rise of resource nationalism among the producer states, the successful push by producers to renegotiate their share of oil revenues, the British withdrawal from the Persian Gulf, and stagnating U.S. production had already put upward pressure on world market oil prices.

Economists also note that counterproductive U.S. energy policies during the crisis contributed to further driving up prices. The U.S. government put price ceilings on "old" (already discovered) oil, while the price of "new" oil (i.e., newly developed reserves and imports) was decontrolled. As a result, some old oil was withdrawn from the market, further pushing up prices. The U.S. government also rationed gasoline, which led to long lines at filling stations. Other countries subject to the embargo that let prices rise, such as the Netherlands, did not face lines. Ironically, long lines at gas stations, the experience most closely associated with the embargo, were caused by poorly thought-out U.S. domestic policies, not the actions of Arab oil producers.

The embargo did result in the adoption of several U.S. and international policy measures to reduce Western economic vulnerabilities to another sudden, sharp reduction in the global supply of oil. Following the embargo, the United States and the other member countries of the OECD established the IEA. The IEA was charged with setting up procedures to manage short-term supply disruptions and to counteract and prevent future disruptions. Members committed to establishing oil reserves equivalent to 90 days of consumption satisfied through imports. The Carter administration subsequently created the SPR. The U.S. government also created the U.S. Department of Energy and passed a series of laws aimed at decreasing the energy intensity of the U.S. economy and reducing the share of imported oil in U.S. consumption: the Energy Reorganization Act in 1974 (Pub. L. No. 93-438), the Energy Policy and Conservation Act (Pub. L. No. 94-163) and the Energy Conservation and Production Act in 1976 (Pub. L. No. 94-385), and the National Energy Act in 1978 (Pub. L. Nos. 95-617, 95-618, 95-619, 95-620, and 95-621). Among other initiatives, these acts imposed efficiency standards on motor vehicles and provided subsidies for the production of synthetic motor-vehicle fuels from coal and oil shale.

Russian Cutoffs of Oil Supplies

The European former Soviet republics import much of their oil (and gas), mostly from Russia. Ever since the other former Soviet republics have become independent, the Russian government has periodically attempted to use oil (and gas) exports to sway political and commercial decisions in these countries.

Latvia. The Soviet Union exported large amounts of oil through the Latvian republic's Ventspils Nafta Terminals. After Latvia regained its independence in 1991, state-owned Russian companies invested in a new Primorsk terminal near St. Petersburg with the backing of the Russian federal and local governments. After the new terminal was completed in early 2003, the Russian state–controlled oil pipeline company Transneft reduced and then halted oil exports through Ventspils, diverting those shipments to the terminal. Since that time, oil shipped through Ventspils Nafta Terminals is delivered by more-costly rail. The shift in transport mode and loss in business dealt a substantial blow to turnover and profits. Taxes paid by Ventspils Nafta Terminals contributed substantially to state revenues. These have fallen sharply.

Russia's decision to build the Transneft-run Primorsk oil terminals appears to have been driven by the Russian government's desire to confine most oil exports to facilities under Russian control. Once these new facilities were built, Russian commercial interests contributed to the decision to stop shipping through Ventspils.

Lithuania. Lithuania's Mažeikių Nafta owns the only large oil refinery in the Baltic states. Once partially owned by the Russian oil company Yukos, after Yukos was forced into bankruptcy by the Russian government, the Lithuanian government auctioned off Yukos's stake. Concerned about the potential for the Russian government to influence the Lithuanian economy, the Lithuanian government sold Yukos's stake to Poland's PKN ORLEN rather than to Russia's state-controlled Rosneft. In late 2006, after the sale, Russia's Transneft pipeline company ceased deliveries to Mažeikių Nafta. Mažeikių Nafta compensated for the lost Russian crude by importing oil through the Būtingė terminal, although it had to pay a higher price. The loss of Russian oil has clearly led to a deterioration in profitability, contributing to a significant loss by Mažeikių Nafta in 2007, although a fire at the refinery was also a major factor. While Transneft claims that "technical reasons" were responsible for the shutdown of the pipeline supplying Mažeikių Nafta, Lithuanian officials believe that the decision was politically motivated to punish Lithuania for the decision to sell Yukos's stake to PKN ORLEN rather than to Rosneft. Although Russia did not use the cutoff to achieve a specific political goal, the cutoff appears to have been designed to send a message concerning the potential costs of thwarting Russian interests.

Belarus. The Druzhba oil pipeline is the largest pipeline connecting Russia with the EU. It transports about 35 percent of Russia's total crude exports to western Europe and supplies one-eighth of the latter's crude oil imports. It has a capacity of 2 mbd, of which 1.4 to 1.6 mbd go directly to consumers in the EU; the remainder is purchased by transit countries—most notably, Belarus, which refines the crude and exports much of the output to Europe.

In early 2007, Russia halted crude oil exports through the Druzhba because of a dispute with Belarus over tariffs and transit fees. At the heart of the dispute stood Moscow's decision to abandon its 16-year-old policy of subsidizing oil export sales to Belarus. Prior to January 2007, Russian exports to Belarus were not subject to Russian export taxes. Since that time, Belarusian importers have to pay the tax, like every other country to which Russia exports oil. Belarus responded by imposing a $45-per-ton additional transit fee ($6.16 per barrel) on Russian oil flowing through the Druzhba pipeline. Belarus also illegally siphoned off oil from the Druzhba destined for other customers. In response, Russia's Transneft stopped shipping oil through the pipeline. The pipeline was reopened after two days, when an agreement on the disputed export duties was reached. Even though most of their oil comes through the Druzhba, refineries in Poland and Germany linked to the pipeline were able to surmount the problems posed by the shutdown. They drew from their inventories to make up for the shortfall, and large refineries in eastern Germany started to arrange for alternative sources of supply in case the disruption stretched out.

While not targeted primarily at western European consumers and even though the interruption was so short that refinery operations were not disrupted, the cutoff was met by a storm of political protest. German chancellor Angela Merkel, who held the EU and Group of Eight (G8) presidencies at the time, called Moscow's move to shut down the pipeline "unacceptable" and said that it risked destroying Russia's credibility as a reliable energy partner. While the dispute was resolved within a relatively short time, the Druzhba incident damaged Russia's reputation as a reliable supplier of energy to the EU, although Russia did succeed in reducing concessions on pricing to Belarus.

Russian Natural-Gas Cutoffs

Although this monograph is focused on imported oil and U.S. national security, in this section, we have also chosen to examine the effectiveness of Russian attempts to use cutoffs in gas supplies to sway decisions by the targeted country. Russia has been embroiled in a series of gas disputes with other former Soviet republics—namely, Georgia, Ukraine, Belarus, and Moldova. Russia either has threatened to or has actually cut off gas supplies to these countries over political as well as commercial disputes. Policymakers and commentators have repeatedly voiced concerns about the potential for Russia to use exports of oil and gas to advance its foreign policy goals. These instances provide additional information on the effectiveness of embargoes to influence other countries.

Gas markets differ from oil markets: Trade in natural gas still takes place in separate regional markets, defined by pipeline networks (see Figure 3.1). In contrast to oil,

Figure 3.1
Russian Oil and Gas Pipelines to Europe

SOURCE: EIA (2007b, p. 11).
RAND *MG838-3.1*

there is not yet a global market for natural gas, although investments in gas-liquefaction facilities, specialized ships, and terminals to offload natural gas are creating the basis for one. Consequently, gas exporters and importers are more closely linked to each other than are their counterparts in oil markets.

Russian Gas Exports. Russia is the largest supplier to and an important transit country for the European gas market. The European market employs long-term bilateral contracts involving offtake agreements and destination clauses. Spot markets play a much smaller role in determining prices and allocating supplies than in the North American market, even though the Eurasian market is considerably larger. In most countries in Europe, the price of natural gas is set with a lag, on the basis of the market prices of a basket of gas substitutes, mostly oil. Responses to shifts in supply and demand take place during negotiations over long-term supply agreements and through smaller spot-market purchases. Gazprom, the state-controlled Russian gas company, responds to differing demand conditions in consuming countries through separate negotiations with each European gas company, many of which are also state controlled. Agreements usually include commitments to take a minimum quantity of natural gas that is stipulated by contract.

In light of this market structure and the high upfront costs for gas pipelines, it becomes very costly for either party to dissolve a contract once a pipeline has been built. This mutual, pipeline-based dependency renders unilateral action by either party difficult. European consumers cannot easily find alternative sources of gas. Gazprom has no other markets to replace Europe. In addition, European markets account for almost all of Gazprom's profits, as the company faces a highly regulated, loss-making domestic market. Curbing supplies to these markets would result in substantial losses. It would also hurt the Russian government's budget revenues because Gazprom has accounted for about 20 percent of overall Russian government revenues in recent years.

Although many of the transit countries through which the large gas pipelines pass are not major gas consumers, Gazprom cannot cut off transit countries without also cutting off downstream deliveries to its larger clients in western Europe.

Since the breakup of the Soviet Union, Russia has sold gas to the other members of the Commonwealth of Independent States (CIS) at significantly lower prices than to its central and western European customers. The Russian government has pressured Gazprom to keep prices low because it desired to preserve former ties and maintain influence and because of its belief that higher prices would exacerbate the transition recessions in its neighbors.[1] Gazprom, for its part, has pursued a strategy of using debts for unpaid gas bills and concessional pricing to acquire local gas transmission and distribution networks. Corruption has also played a role in keeping official prices low: The agreements by which gas is sold to these countries, especially Ukraine, have been murky, involving intermediary companies, resales of gas from central Asia, and mixtures of prices and gas from many sources. These intermediary companies have skimmed substantial sums of money from the gas trade, some of which has reportedly ended up in the pockets of government officials. Ukraine and Belarus have also successfully managed to preserve lower prices by taking advantage of Russia's dependence on their transit pipelines to ship gas to western Europe. Belarus and Armenia were rewarded for adopting pro-Russia foreign policy stances.

In late 2005, Gazprom began to demand higher prices. It first announced a sharp increase in the price of gas sold to Georgia, with which Russia has had poor relations. Shortly thereafter, it increased prices to Ukraine and Moldova. Because relations between the Russian government and the new leaders who had emerged during the color revolutions in Georgia (the Rose Revolution, 2003–2004) and Ukraine (the Orange

---

[1]   Transition recessions are the economic recessions that accompanied the transition from centrally planned to market economies.

Revolution, 2004–2005) were strained, the price increases were perceived in the West and domestically as punishment for the new governments' pro-Western policy stances. While political considerations appear to have played an important role in the timing of the price increases, increasing Gazprom profits from sales of gas to these countries was also a factor. Gazprom appears to have convinced the Russian government that selling gas to other former Soviet republics at prices lower than could be earned elsewhere has not been a useful foreign policy instrument. It also convinced the Russian government that it would be better off if Gazprom earned more money from higher prices. Despite warm relations with Belarus and Armenia, Gazprom raised gas prices for Belarus by 400 percent in late 2006, and Armenia also had to pay more. Prices charged Georgia have been similar to those charged Armenia, in spite of the acrimony between Georgia and Russia. The gas price increases have had a heavy flavor of "business is business" as Gazprom hunts for more revenues and profits.

The price increases have facilitated Gazprom's quest to acquire equity stakes in gas transmission and distribution systems in countries that are its customers. Belarus and Moldova have agreed to sell stakes in their systems to cover the higher costs of the gas. These acquisitions strengthen Gazprom's control over transit pipelines and cut down on the number of profit-seeking intermediaries that demand a cut from the gas trade in exchange for transit rights.

Ukraine. In December 2006, after Ukraine refused to pay the higher prices that Gazprom demanded, Gazprom informed Ukraine that it would no longer supply it with gas, reducing the volume of gas transiting Ukrainian pipelines by the amount that Ukraine consumes. Ukraine responded by siphoning off gas that had been sold to western European customers for its own consumption, while insisting that it would pay the former, lower price. Gazprom responded by cutting off all gas. To the Russian government's surprise, western European governments and customers harshly criticized Russia for the cutoff, ignoring the Russian government's arguments that the real culprit was Ukraine. Gazprom restarted gas flows almost immediately. Russia and Ukraine then negotiated a complex compromise involving mixing cheaper central Asian gas with more-expensive Russian gas that was then sold to Ukraine at a fixed average price substantially below the prices charged western European customers. As part of the deal, Russia also agreed to pay Ukraine higher transit fees.

Ukraine and Gazprom have continued to argue about price increases and charges. In March 2008, Gazprom threatened to cut off gas supplies over disagreements over payments for previous gas deliveries. Ukrainian officials openly warned that they would again siphon off gas intended for western Europe to compensate for any reductions in supplies by Gazprom. In January 2009, Gazprom followed through on its threat, cutting off gas to Ukraine and, eventually, the rest of Europe.

The Ukrainian government believed that the cutoffs were inspired by political as well as economic reasons. The Russian government had actively attempted to thwart the Orange Revolution, which resulted in a second election for the presidency, which

Russia's preferred candidate lost. Because of concerns about potential Russian economic leverage, Ukrainian governments have opposed Russian attempts to acquire parts of its gas transmission and distribution networks. The Ukrainian government passed a law in 2008 forbidding the sale of these assets to Russia.

Georgia. During the winter of 2005–2006, Gazprom claimed that the gas transmission line to Georgia had been sabotaged, resulting in a cutoff in gas supplies to Georgia. The Georgian government, which assumed that the cutoff was deliberate, responded by expanding links between its gas distribution system and Azerbaijan, diversifying sources of supply. The cutoff intensified the Georgian government's suspicions concerning Russia. It failed to sway Georgian domestic policies and has left Georgia less dependent on Russian gas.

Belarus. In contrast to Ukraine, Belarus opted to sell Gazprom a majority stake in Beltransgaz, the state-controlled company that owns Belarus's gas network. Gazprom compensated Belarus for the network by slowing the rate of price increases. However, gas prices will still be steadily increased and are scheduled to reach western European levels by 2011. Higher gas prices have apparently hit the Belarusian economy hard. The Lukashenko government has been looking for loans to cover the increased cost of imported gas.

Moldova. In January 2006, Gazprom cut off gas supplies to Moldova after the Moldovan government refused to accept a 100-percent increase in gas prices. Gazprom restored supplies later that month, after Moldova negotiated a somewhat smaller increase. To raise money to cover the increased costs of the gas and as a bargaining chip to moderate the increase, the Moldovan government agreed to sell an additional share of the national gas company MoldovaGaz to Gazprom. Gazprom was already the majority stakeholder in the company.

Over the course of the past several years, Gazprom has succeeded in achieving most of its implicit corporate goals in the CIS. It has substantially narrowed the gap between CIS and western European prices after the costs of transit charges paid by the western Europeans are taken into account. It has negotiated price agreements to eliminate the gap by 2011. It has acquired substantial stakes in gas pipeline networks in Belarus and Moldova.

The Russian government has been less successful in using threats of gas cutoffs to pursue foreign policy goals. Recent elections in Georgia and Ukraine have kept in or returned to office leaders who have pursued foreign policies independent of Russia. Georgia has diversified gas supplied to Azerbaijan, reducing its dependence on Russia for gas. The price increases have also weakened the Russian government's ability to influence these countries, as gas prices are increasingly determined by market forces, not political bargains.

Oil Export Subsidies

The historical record indicates that the oil (and gas) "stick," although frequently economically disruptive, has not been particularly successful at influencing governments to shift policies, especially on issues of national importance. What, then, of energy "carrots"? A number of oil and gas producers have sold oil and gas to favored customers at a discount from the prices they charge less favored customers. Others have provided foreign assistance financed by sales of oil. In some instances, price breaks or grants have been provided for humanitarian reasons, but often they have been provided to curry favor or support weak allies. By providing a subsidy or assistance, energy exporters hope to sway the recipients to act in ways more to their liking. In this section, we assess the efficacy of these policies.

Soviet Subsidies to Eastern Europe

Price discounts do not appear to have purchased much in the way of loyalty. The Soviet Union provided the other members of the Warsaw Pact (Warsaw Treaty of Friendship, Cooperation and Mutual Assistance) and Cuba with massive subsidies between 1975 and 1983 in the form of below-market prices for oil and natural gas. These subsidies were estimated at $19 billion for 1981 alone (Marrese and Vanous, 1983). Because of the opaque pricing and trade arrangements within the Council for Mutual Economic Assistance, the Soviet counterpart to the Common Market (the European Economic Community, or EEC), neither the Soviet Union nor the eastern European recipients were sure of the size of the transfers. Once the central and eastern European states overthrew their communist leaderships, the Soviet Union rapidly phased out sales of energy at favorable prices. Few in central and eastern Europe rued the change in the relationship, despite the loss of subsidies.

Russian Subsidies to Other Members of the CIS

After the Soviet Union collapsed, Russia was slow to raise energy prices for the other members of the CIS. Initially, both oil and gas were sold at lower prices than those available to western or central European buyers. However, by the mid-1990s, most Russian oil was being sold to other members of the CIS at market prices. The exception was Belarus. As noted earlier, until 2007, Belarusian purchasers of Russian oil did not have to pay Russian export taxes or other customs charges, giving Belarus a substantial discount. Belarusian refineries processed the cheap Russian oil and resold petroleum products at market prices to western Europe and elsewhere. Fuels and petrochemicals accounted for 40 percent of Belarus's exports. Through its subsidy policy, Russia helped the authoritarian Lukashenko government remain in power despite Belarus's poor economic performance, and Russia kept a committed ally. When Russia put a halt to the subsidies, the two governments quarreled, but Belarus did not stop supporting Russia's foreign policy positions.

Natural gas continues to be sold to members of the CIS at more-favorable prices than to western Europe. Even at relatively low prices, during the 1990s, all the energy-importing countries of the CIS racked up substantial debts with Russian energy companies and, through them, the Russian government. These debts have been reduced and rescheduled through a combination of transfers of assets, primarily in the energy sector, and debt forgiveness. Through these settlements, Gazprom acquired substantial stakes in gas transmission and distribution networks in these countries. However, in the cases of Georgia and Ukraine, the subsidies have not bought enduring support for Russian policies.

Assistance to Egypt from the Persian Gulf States

Since the 1950s, foreign assistance has played an important role in financing the Egyptian government and supporting the Egyptian economy. At the Khartoum Resolution conference in September 1967 following the June 1967 Six-Day War, Arab oil states committed themselves to providing substantial assistance to Egypt to help it contend with Israel. According to former Egyptian prime minister Kamal Hassan Ali, Egypt received around $5 billion from the Arab states between 1973 and 1977. According to donors, Egypt received as much as $17 billion in this time frame (Feiler, 2003, p. 1).

In retaliation for signing the Camp David Accords on September 17, 1978, Egypt's Arab donors imposed political and economic sanctions against Egypt. They cut off all foreign aid, severed diplomatic relations, and refused to sell oil to Egypt. The headquarters of the Arab League (the League of Arab States) was moved from Cairo to Tunisia.

Egypt's decision to sign the Camp David Accords is instructive. It chose to give up subsidized oil imports, billions of dollars in aid, and the approval of the Arab community in exchange for Israeli concessions affecting its territorial integrity and national security: the return of the Sinai, a peace agreement with Israel, and a close relationship with the United States, including substantial U.S. economic aid. The threatened loss of subsidized oil from Arab oil-exporting states failed to dissuade Egypt from signing the accords.

Iraqi Subsidies to Syria and Jordan

The Kirkuk-Banias pipeline was built in the late 1940s to transship oil from Iraq's Kirkuk oil fields to Banias, Syria, and Tripoli, Lebanon. During the Iran-Iraq War, Syria, an ally of Iran, chose to shut this pipeline, forgoing transit fees. The shutdown deprived Iraq of half its remaining oil export capacity. Syria's political calculations overrode its economic interests.

In contrast, despite pressure from the international community, Jordan remained officially neutral during the First Gulf War, although it observed the United Nations (UN) embargo and helped the allied effort in other ways. Throughout the 1980s, Iraq had satisfied most of Jordan's energy needs through the sale of oil at below-market prices. After its defeat, Iraq continued to supply Jordan with oil at below-market prices in the context of the UN's Oil-for-Food Programme. Although Jordanian policy-makers recognized the benefits that Jordan received from favorable prices for oil before and after the war, widespread support for Iraq among Jordan's large Palestinian community was probably a more important factor in the Jordanian government's decision to maintain relations with Iraq.

Below-Market Sales of Oil by Venezuela and Saudi Arabia

The government of Venezuela has sold oil at discounted prices to Bolivia. Saudi Arabia has sold oil to Yemen at discounted prices. Periodically, many members of OPEC have given away or sold oil at discounted prices to favored states.

The Bolivian government has supported Venezuela and its president, Hugo Chavez, in international forums. Yemen and Jordan have supported Saudi Arabian positions. However, Kuwaiti gifts have not always translated into support. After it was invaded by Iraq, Yemen and the Palestinian authorities supported Iraq, despite having received substantial assistance from the Kuwaiti government in the past. Jordan uneasily sat on the fence. In the oil trade, carrots appear to be modestly more effective than sticks. However, they have had mixed success in inducing support: Many recipients have been fair-weather friends.

Securing Oil Supplies

Some governments have attempted to secure sources of oil supply by investing in or otherwise establishing special relationships with oil exporters. Quests to create special arrangements might be detrimental to U.S. national security for the following reasons:

- They may undercut the operations of the global oil market, making it less resilient, thereby reducing the ability of the United States to tap alternative sources of supply, if there were to be an abrupt decline in output.
- Consuming countries that seek to make these investments may tailor their foreign policies to align with oil exporters rather than pursue common goals with the United States.

China, India, and Japan, among other countries, have attempted to secure sources of supply by diplomatic means, equity investments, subsidized loans, and development aid. This section assesses the costs, successes, failures, and political consequences of attempts by two consuming nations—China and Japan, the two largest oil importers after the United States—to lock up sources of imported oil.

China

China switched from being a major oil exporter in the 1970s and 1980s to being a net importer of oil in 1993. It is currently the second largest consumer of oil in the world after the United States, importing more than half of its consumption (BP, 2007). Most of China's producing oil fields have matured; many are in decline. China's crude oil net imports are projected to continue to rise sharply (see IEA, 2007, p. 124).

The Chinese government has sought to secure foreign supplies of oil through its national oil companies (NOCs): China National Petroleum Corporation (CNPC), China National Offshore Oil Corporation (CNOOC), and China Petroleum and Chemical Corporation (Sinopec). It first turned to smaller oil producers, such as Oman, seeking exclusive drilling rights to secure oil supplies. More recently, China has sought oil concessions in Africa, Central Asia, and Iran.

The Chinese government has supported the efforts of these companies to secure preferential deals both diplomatically and by offering the host countries Chinese government loans.[2] China has provided military assistance and development aid to oil exporters, most recently through its $5 billion China-Africa Development Fund.

China's energy diplomacy has had mixed success. Chinese companies sought to secure supplies of oil during a period of record-high oil prices (DOE, 2006, p. 32). China has tried to avoid tensions with the United States by avoiding oil-producing regions with a strong U.S. corporate, diplomatic, or military presence. China's NOCs have generally targeted smaller fields or less stable countries, where the political and economic risks are higher. Although terms are often not made publicly available, in some instances, Chinese oil companies appear to have received less favorable terms than those the major international oil companies find necessary to generate an adequate rate of return.

Chinese investments in foreign oil assets are not the major component of Chinese foreign direct investment (FDI) abroad. The Eurasia Group estimates that 12 percent of total Chinese outward FDI in 2005 was directed at the oil, gas, and mining sectors. Chinese FDI in oil, gas, and mining ran $1.38 billion in 2003, $1.8 billion in 2004, and $1.7 billion in 2005 (Eurasia Group, 2006, p. 5). These flows have been more or less stable, while FDI in other sectors has been rapidly growing. Chinese total outward FDI rose from $4.2 billion in 2003 to $14 billion in 2005.

What has this effort bought? China's NOCs have taken equity stakes in a number of new oil developments. However, to date, Chinese FDI in oil projects has been relatively small. The only two projects in which Chinese oil companies own majority stakes and that export more than 100,000 barrels per day (bpd) to China are Aktobemunaigaz in Kazakhstan and Sudan's Greater Nile Petroleum Operating Company (GNPOC) (Eurasia Group, 2006, p. 4). Most of the oil from these projects is sold

---

[2]   On the causes of and changes in China's "going-out" strategy, see House (2008, p. 160). See also Xiaojie (2007).

on the open market to whichever buyer is willing to pay the highest price. In 2005, exports from Chinese-owned foreign projects covered 10 and 15 percent of Chinese oil consumption.[3] The projects have increased world oil supplies, easing supply pressures on the global market. To the extent that China's diplomatic and foreign assistance initiatives in oil-exporting countries are designed solely to induce local governments to give concessions to Chinese NOCs, they add to the cost of Chinese imports of oil. However, as most governments, including the Chinese, simultaneously pursue more than one foreign policy goal, these costs of diplomacy should not all be ascribed to the pursuit of oil. Some of these initiatives have cost China in terms of reputation. China's reluctance to permit UN sanctions against Sudan because of the Sudanese government's behavior in the Darfur region has been condemned by human-rights groups and criticized by a number of Western governments.

In short, China's policy of encouraging its NOCs to invest in oil projects outside of China has marginally increased world market oil supplies. Because most of the oil from these projects is sold to third-party countries, these ventures have not contributed significantly to improving China's physical security of supply. Some of these ventures—most notably, projects in Sudan—have come at some cost to China's reputation as a responsible member of the international community.

Japan's Energy Diplomacy

Japan imports virtually all of its oil. In contrast to China, private, not state-controlled, companies account for the bulk of overseas exploration and production projects in which Japanese companies are involved. Currently, 70 private Japanese companies are engaged in exploration and production abroad, the largest of which are INPEX Corporation, Japan Petroleum Exploration Corporation (JAPEX), and the Arabian Oil Company (AOC) (Paik et al., 2007).

Predating China's current policies, the Japanese government took an active role in promoting Japanese investment in upstream oil and gas projects. Prior to the 1970s, state-owned Japan National Oil Corporation (JNOC) was tasked with making equity investments in international oil projects. In the 1970s and 1980s, the government of Japan scaled back this policy (see, e.g., Ziegler, 2008). Oil importers were free to purchase oil at the most favorable prices available.

China's policy of encouraging its state-controlled oil companies to invest in projects in developing countries appears to have led to a reassessment in Japan. The Japanese government has recently taken a more-active role in encouraging Japanese companies to invest in new projects abroad. This more-assertive role of the government in securing Japan's energy supplies is reflected in Japan's New National Energy Strategy, released by Japan's Ministry of Economy, Trade, and Industry (METI) in 2006 (Evans, 2006,

---

[3] See Eurasia Group (2006, p. 3) and DOE (2006, p. 28). For an overview of Chinese foreign assets, see also KPMG (2005) and Downs (2005).

pp. 8–9). Besides setting clear targets for increasing energy efficiency and diversifying the energy mix away from oil, the strategy makes a central goal of increasing the percentage of Japanese oil imports coming from projects in which Japanese companies hold equity stakes. The share of this oil in total crude oil imports is supposed to rise from the 2007 level of 15 percent to 40 percent by 2030.

The Japanese government supports overseas oil ventures by fronting risk money and offering loans on favorable terms for exploration and production projects, reducing the costs to Japanese companies, thereby making it possible for them to submit lower bids than could companies from countries that do not provide these subsidies. The Japanese government also offers risk insurance. It supports bids by strengthening diplomatic relations with oil-rich countries and by targeting development aid to oil producers (Paik et al., 2007, pp. 21–22). In Central Asia, Japan launched the Silk Road Energy Mission to Kazakhstan, Uzbekistan, Azerbaijan, and Turkmenistan in 2002 and has intensified bilateral relations with the region.

As in China's case, it is questionable how successful the Japanese government has been or is likely to be in improving its access to oil through this policy. Raising the share of imported oil to 40 percent from ventures in which Japanese firms hold equity by 2030 means that Japanese companies will have to double output from equity ventures by that time (Paik et al., 2007, p. 23). To reach this target, Japanese oil companies will probably have to invest in considerably riskier projects or offer more-attractive terms to partners than otherwise would be the case. In some instances, they may be induced to invest in loss-making projects. Peter Evans, an analyst with Cambridge Energy Research Associates, notes, "Japan has a record of spending lavishly on projects that do not significantly increase the country's energy security" (Evans, 2006, p. 20). In the late 1990s, almost all of JNOC's subsidiaries and foreign ventures were losing money. Because of this poor performance, JNOC was closed and replaced by the Japan Oil, Gas, and Metals National Corporation (JOGMEC). Japan's current portfolio of upstream projects is not of the best quality, although a consortium of Japanese energy firms won licenses for promising exploration and production projects in Libya in 2005. Japan has recently suffered some major setbacks in Iran's Azadegan oil field and the Khafji oil field in the neutral zone between Kuwait and Saudi Arabia (Mitsumori, 2003). Japanese projects in Eurasia tend to be small, with the exception of Sakhalin-I, in which JOGMEC provided loan guarantees for Sakhalin Oil and Gas Development Company's (SODECO's) 30-percent stake. While the share of Japanese oil imports from Africa is currently rising, equity oil from the continent is negligible, despite increased diplomatic and financial efforts by the Japanese government (Masaki, 2007; Ziegler, 2008, p. 144).

The major risk in Japan's strategy of resorting to state-backed upstream deals lies less in actual volumes than that government-induced competition for equity stakes will drive up the cost of investing in projects and encourage Japanese oil companies to take undue risks. Japan's example could also trigger similar actions by other consuming nations, exacerbating competition for energy supplies and potentially fragmenting international oil markets. Japanese policies could weaken OECD efforts to commit consuming nations to rely on internationally integrated oil markets to ensure energy security.

Conclusions

The historical record indicates that neither oil nor gas embargoes have been particularly successful as foreign policy instruments. Prior to World War II, oil embargoes either were circumvented or may have served as a trigger to conflict. The Arab oil embargoes, especially the 1973–1974 embargo and the associated 25-percent reduction in oil output, hurt the global economy. However, because international oil markets are integrated, producers cannot target an individual country. Oil flows to the highest bidder, so reductions in output push up world market prices, affecting all oil-importing countries. Russia's attempts to use natural-gas exports as a foreign policy tool have also not met with much success.

This is not to say that both buyers and sellers of oil do not attempt to maintain amicable relationships. In most business relationships, customers and suppliers prefer to stay on friendly terms. The U.S. government listens to Saudi Arabia; the German government attempts to maintain friendly relations with Russia. However, national interests have trumped efforts by suppliers to wield influence. In fact, what is striking about the global oil market is how acrimonious political relations often are between buyers and sellers. Mexico supplied the United States with oil during decades when relations were cold. Venezuela continues to sell oil to the United States despite the antipathy that exists between the two countries' governments.

Providing oil carrots instead of using the embargo stick has been somewhat more successful. Providing oil at subsidized prices may not buy an exporter love, but it can induce beneficiaries to support some of the donor's issues in international forums. However, beneficiaries have had short memories and have not supported policies contrary to their national interests. A number of recipients of oil-financed assistance from Kuwait did not support the international effort to expel Saddam Hussein's forces. Russia has not gained much support from neighbors to which it has provided natural gas sold at discounted prices. Venezuela's grants of discounted fuel oil to poor U.S. citizens have not resulted in an improvement in U.S.-Venezuela relations.

Governments' efforts to secure supplies of oil by supporting investments by national firms in exploration and production projects in other countries have not been notably successful. Most of these projects sell oil to the highest bidder, not to the investing country. Sales are made at market prices. As long as price determines who gets oil, equity stakes do not buy much in the way of increased security of supply.

CHAPTER FOUR

Oil Revenues, Rogue States, and Terrorist Groups

Oil Revenues and Rogue States

Many U.S. citizens and policymakers are concerned that proceeds from payments for imported oil are being used to finance activities contrary to U.S. interests. Governments of some countries openly hostile to the United States—Iran and Venezuela, in particular—rely on oil exports for most of their budget revenues. To the extent that global consumption of oil contributes to increasing the revenues of these governments, directly or through the effects of U.S. consumption on global demand for oil, global importers' payments for imported oil may help finance governments intent on thwarting U.S. policies.

Oil exports are not a necessary condition for financing rogue states. North Korea is an oil importer but has built nuclear weapons. In the 1990s, when under the rule of the Taliban, Afghanistan, another oil importer, became a sanctuary for al Qaeda. In the 1990s, before it began exporting oil in large quantities, Sudan harbored Osama bin Laden.

Most major oil exporters—for example, Canada, the largest supplier of oil to the United States—are not hostile to the United States. However, in recent decades, a few oil exporters have threatened U.S. interests and, in some cases, global peace. Between 1980 and 1990, Iraq invaded two of its neighbors, Iran and Kuwait. Iran has perpetrated terrorist attacks on U.S. installations, financed the activities of Hizballah, and supported violent militias and insurgents in Iraq and Afghanistan and is engaged in a nuclear program that is likely to produce enough enriched uranium to manufacture nuclear weapons. Although the hostile activities of Venezuela's government under Hugo Chavez are not on the order of Iran's, Venezuela is pursuing a foreign policy designed to thwart U.S. policy goals in Latin America.

The ability of these governments to pursue policies contrary to U.S. interests depends, in part, on financial resources. Because oil exports are such important sources of revenues for the budgets of Iran and Venezuela, if budget revenues were lower, either because of lower prices or smaller export volumes, the ability of these governments to oppose U.S. interests would be impaired. In this section, we evaluate the role of oil exports in government revenues, the successes of initiatives these governments have pursued to thwart U.S. foreign policies, and the limitations these governments face in maintaining or expanding these initiatives.

Iran

Iran's Oil Revenues. In the 1990s and the first part of this decade, Iran's oil production recovered from the effects of the 1979 Iranian Revolution and the Iran-Iraq War, peaking in 2005 (Figure 4.1). Iran now accounts for 4.8 percent of global production. Despite the many inefficiencies of the National Iranian Oil Company (NIOC), a combination of increased investment, better-trained staff, and more-effective use of foreign contractors has allowed NIOC to shore up production.

Oil export revenues are crucial to the Iranian government's finances, accounting for 70 percent of total budget revenues in 2007. Net oil exports also rose in the 1990s and the first part of this decade. However, they have fallen off since 2005 because of the decline in production and because an increasing share of output is being refined for domestic use.[1]

Figure 4.1
Iranian Oil Production, Exports, and Domestic Consumption

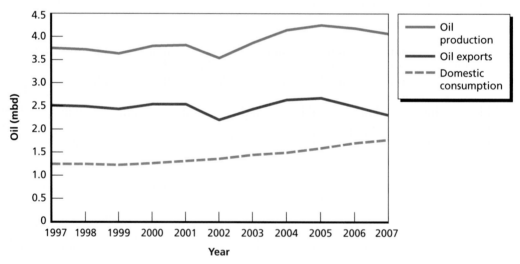

SOURCE: EIA (undated).
RAND MG838-4.1

---

[1]   The state-owned NIOC subsidizes domestic purchases of refined oil products by refining and selling fuels on the domestic market at less than their market value and by paying for imports of gasoline and other fuels that exceed the capacity of the domestic refining industry. Because of price controls, NIOC and, hence, the government lose money on domestic sales of these oil products. The costs of these fuels and the associated revenue losses suffered by NIOC because of subsidizing these products are a major expense for the Iranian government, running several percentage points of Iran's GDP.

Over the course of the past decade, the Iranian government has enjoyed a bonanza in oil revenues (Figure 4.2). Despite the recent decline in Iranian export volumes, higher world market oil prices boosted government revenues in 2007 and the first part of 2008. Declines in world market oil prices in the second half of 2008 reduced revenues and severely threatened the budget in 2009.

The Iranian government has used these revenues to increase total government spending from $13 billion in 1999 to $70 billion in 2008. The increases have gone toward higher salaries for government workers, subsidies—especially subsidies for gasoline and diesel fuel—public investment, and other activities, including military procurement, support for Hizballah, and the development of Iran's nuclear program. The sharp decline in oil prices in the second half of 2008 and early 2009 is forcing the Iranian government to slash expenditures on some of these items.

Iranian Policies Contrary to U.S. Interests

Nuclear Program. Iran appears to be on its way to becoming a nuclear power. The Iranian government claims that its nuclear program is designed to generate electricity. However, Iran's failure to fully comply with UN Security Council and International Atomic Energy Agency resolutions calling for the cessation of uranium enrichment and its large untapped reserves of natural gas, a fuel well-suited for generating electricity, have led many analysts and policymakers to conclude that the nuclear program is geared toward acquiring nuclear weapons.

Based on the costs of nuclear programs in other countries, Iran's program must have run several billion dollars over the past few decades. Iran has built an

Figure 4.2
Iranian Budget Revenues and Expenditures

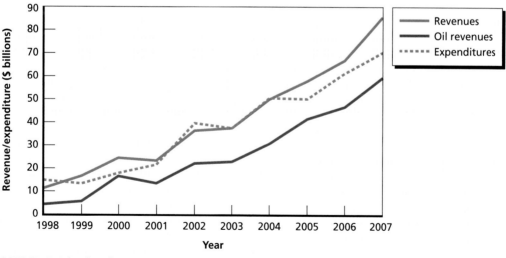

SOURCE: IMF (undated).
RAND MG838-4.2

underground uranium-enrichment facility in Natanz, a uranium hexafluoride–conversion plant in Eṣfahān, a nuclear reactor in Būshehr, and a heavy water plant in Arak. Iran is also believed to operate dozens of additional nuclear sites, including research laboratories, centrifuge-manufacturing sites, and, most likely, secret military installations related to the weaponization of the nuclear program.

Iran's increased oil revenues have enabled it to allocate additional resources for the construction and operation of its nuclear facilities. They have also enabled it to more easily weather international sanctions meant to slow or stop its uranium-enrichment activities. Sanctions have raised the costs of doing business in Iran by making it more difficult to obtain letters of credit. The Iranian government has stepped in, providing credit for investments and imports, credit made possible by these higher oil revenues.

Military Expenditures. Higher oil revenues have allowed Iran to increase expenditures on military programs. Military spending, estimated at $1.77 billion in 1998, has risen to an estimated $8.4 billion in 2008 (IISS, various years). Iran has used some of these funds to invest in a domestic missile program, purchasing technologies, components, and technical assistance from foreign suppliers, including North Korea, Russia, and China. Iran's medium-range missiles could be used to carry nuclear weapons and have a long enough range to reach Israel and U.S. forces stationed in the Persian Gulf. Iran has also invested in a nascent space program. The Safir missile, which can be used to launch satellites, was tested in August 2008. Technologies used for the Safir are needed to develop an intercontinental ballistic missile.

Iran's military leaders realize that Iran's aging conventional navy is no match for the U.S. 5th Fleet. To counter U.S. naval superiority, Iran has invested heavily in developing asymmetrical naval capabilities, including fast attack boats designed to swarm U.S. ships; relatively advanced antiship cruise missiles; midget submarines; and mines that can be deployed along Persian Gulf shipping routes. Iran's asymmetrical naval capabilities could be used to disrupt shipping in the Persian Gulf, especially in the event of a military conflict with the United States.

Iraq. Iran has used its financial resources to enhance its influence in Iraq, particularly in the predominantly Shi'a southern region. The U.S. military has accused the Islamic Revolutionary Guard Corps' (IRGC's) specialized Qods Force of providing financial and military support to Shi'a political groups, especially the Jaish al Mahdi (JAM), that oppose the U.S. presence in Iraq. The Qods Force is believed to be the source of explosively formed projectiles, a highly lethal type of roadside bomb that has killed large numbers of U.S. and coalition soldiers and Iraqis. Iran is believed to be training JAM militia members in camps in Iran. According to some estimates, Iranian funding for JAM may reach tens of millions of dollars each month.

Iran has spent considerable sums to enhance its influence in Iraq by funding the construction of Shi'a cultural centers and donates to Islamic charities, including refurbishing the holy shrines in An Najaf (Kemp, 2005). Iran has funded religious training for Iraqi religious seminarians. If it so desires, Iran might be able to use its ties with

JAM and JAM's soft power in Iraq to launch proxy attacks against U.S. forces in that country by groups that it has cultivated.

Palestinian-Israeli Conflict. Iran has become an important military, political, and religious supporter of the Lebanese Shi'a group, Hizballah. Iran has provided financial assistance to Hizballah since it emerged in the early 1980s. Hizballah is considered Iran's key military proxy in the Arab world. Hizballah's forces have been used to retaliate against Israel and the United States and could do so again in the event of military strikes on Iran's nuclear infrastructure. Iran provides much of Hizballah's funding, training, and military supplies, including thousands of short-range missiles, more-sophisticated antitank guided missiles, and antiship cruise missiles.

Hizballah took advantage of its military capabilities, largely developed with Iranian assistance, in its 2006 war with Israel, sparked when Hizballah kidnapped two Israeli soldiers. During the war, Hizballah (and Lebanon) sustained heavy casualties and damage from Israeli forces, but Hizballah was able to withstand the Israeli military assault. Throughout the conflict, Hizballah was able to launch large numbers of missiles into northern Israel. By depicting its resistance to Israel as a military and psychological victory, Hizballah was able to enhance its image throughout much of the Middle East.

Hizballah's social services have fostered support for Iran within a large segment of Lebanon's Shi'a community. Neglected by the central government in Beirut, pro-Hizballah Shi'as rely on Hizballah for educational, medical, and religious services. In addition to its military support, Iran is believed to provide much of the funding for Hizballah's social-welfare activities. Total Iranian funding for Hizballah (military and social) is estimated to be as high as $200 million per year (Wilson, 2004).

Hamas also receives substantial funding from Iran, especially since its victory in the Palestinian elections in 2006. After the elections, Iran stepped in to fill the vacuum created by the West's reluctance to fund the Hamas-dominated Palestinian National Authority (PNA). According to one estimate, Iran may have promised Hamas up to $250 million in 2006. Some analysts viewed Hamas' takeover of the Gaza Strip in June 2007 as a strategic victory. Iran is also believed to be training Hamas fighters, who have been firing rockets into Israel from the Gaza Strip, fueling the recent conflict.

Oil as a Political Weapon. Iran has periodically threatened to use oil as a strategic and diplomatic instrument. One of Iran's top nuclear negotiators, Javad Vaeedi, warned in 2006 that Iran would cut off its oil exports in order to drive up global prices in the face of threats to its nuclear program.

Iran has also repeatedly threatened to close off the Strait of Hormuz in the event of an attack on its nuclear facilities. IRGC commander-in-chief Mohammad Ali Jafari has stated, "Enemies know that we are easily able to block the Strait of Hormuz for an unlimited period" (Lake, 2008). If Iran successfully disrupted oil traffic through the strait, world market oil prices would rise sharply (Evans-Pritchard, 2008).

Limitations. The Iranian government's threats to reduce oil exports have been empty because it depends so heavily on them to finance its expenditures. If oil prices were to rise, Iran might be tempted to announce a small cutback to increase prices, especially if the Ahmad-Nejad government believes that it would recoup more from higher prices than it would lose through lower export volumes. However, a large reduction in oil export volumes would result in severe financial and balance-of-payments difficulties for the Iranian government, especially as foreign currency reserves decline.[2]

Iran's ability to close the strait indefinitely is doubtful. Iranian naval forces are less capable than the U.S. 5th Fleet, stationed in Bahrain. In addition to the likely military response, an Iranian attempt to close the strait would result in severe financial hardship for the Iranian government because of the fall in its own exports. An attempt to block the strait would severely damage its relations with the Cooperation Council for the Arab States of the Gulf (Gulf Cooperation Council, or GCC) states. Iran is heavily dependent on its commercial relations with the United Arab Emirates, especially Dubai, and cannot afford to rupture its relations with that emirate.

Venezuela

Venezuela's Oil Revenues. In 2007, Venezuela produced 2.667 million barrels of oil and associated liquids per day. Of this, 1.93 mbd were exported and approximately 0.74 mbd were allocated to domestic consumption at subsidized prices. Because the Venezuelan government sells gasoline at $0.12 per gallon, smuggling gasoline to neighboring states, such as Colombia, is very profitable; domestic consumption includes these unofficial exports.

Venezuela earned well over $60 billion from exports of petroleum and refined oil products in 2007, of which $51.6 billion consisted of exports of crude oil (from IMF data). Despite the antipathy between the U.S. and Venezuelan governments, the United States remains Venezuela's biggest customer. In 2007, the United States purchased $38.8 billion of petroleum and refined oil products from Venezuela. In that same year, the United States purchased 64 percent of Venezuela's total exports of oil (from UN data provided in 2009). Commercial relations will be hard to break. Because of Venezuela's proximity to the United States, transport costs are much lower to U.S. markets than to other major oil importers, such as the EU, China, or Japan. Because of these long-standing supply relations, U.S. oil companies have built and designed certain refineries to process Venezuela's heavy crude oils. For its part, in 1986, the Venezuelan government purchased half of CITGO, a major U.S. oil refiner and distributor, to ensure a market for its heavy crude oil; it purchased the remainder in 1990. Because some U.S. refiners have made large investments in equipment to process Venezuela's heavy crudes, these refiners are able to purchase these crudes at a discount because less sophisticated refineries have difficulty in processing them. For example,

---

[2]   In the first half of 2008, foreign currency reserves were large enough to cover 12 months of imports.

Venezuela's Bachaquero 17 has been sold at more than a 25-percent discount to West Texas Intermediate.

The Venezuelan government relied on revenues from oil for 53 percent of government revenues in 2006 (Figure 4.3). In current dollars, government revenues from oil have risen from $5.3 billion in 1998, the bottom of the oil bust in the 1990s, to $29.3 billion in 2006. Revenues surged again in 2007 and the first half of 2008, but the Venezuelan government has had to sharply cut spending in 2009 as oil prices have plummeted.

Oil Exports and Venezuela's Policies Contrary to U.S. Interests. Venezuela's president, Hugo Chavez, has used some of these revenues to consolidate his political base, expand Venezuela's influence throughout Latin America and the Caribbean, and build up Venezuela's military forces. Chavez has also attempted to create coalitions of countries to counterbalance U.S. international influence—most notably, with Iran but also, less successfully, with China and Russia.

Consolidating the Political Base. Most of the increases in Venezuelan government revenues stemming from higher oil prices between 2007 and the first half of 2008 have gone for social spending. Chavez has cultivated lower-income voters by increasing spending on social programs to improve education and health care and to subsidize food and energy. Social spending has risen from $7.5 billion in 1998 to $25.1 billion in 2006; $38.6 billion if social spending by Petróleos de Venezuela S.A. (PDVSA), Venezuela's state-owned oil company, is included. Higher revenues from oil have permitted

Figure 4.3
Oil Exports as a Percentage of GDP and Oil Revenues as a Percentage of Total Government Revenues in Venezuela

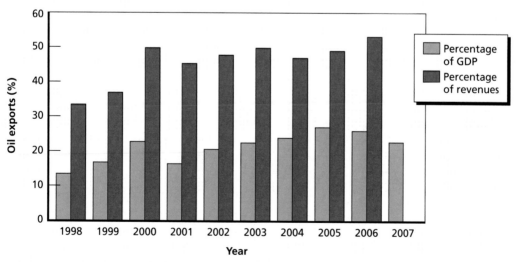

SOURCES: Calculated from IMF (undated) and Weisbrot and Sandoval (2008).
RAND *MG838-4.3*

Chavez to spend three to five times more on social spending in dollar terms over the past decade, helping him to keep the support of his core political constituency, Venezuela's poor. Chavez has also nationalized companies and renegotiated the terms of projects—most notably, the stakes of international companies in new oil fields. Large parts of Venezuela's economy have come under government control during his presidency.

Building Influence in Latin America. The Venezuelan government has funded political parties and presidential candidates in Bolivia, Ecuador, Nicaragua, Paraguay, and Peru. All the candidates backed by Chavez have won in recent elections except Ollanta Moisés Humala Tasso, who was defeated by Alan García in Peru's 2006 presidential election. Humala, like Chavez, is a former military officer who led an unsuccessful uprising against his government. All the elections were judged free and fair; domestic considerations drove the results. However, Chavez's financial support was welcomed. Leaders he has supported have come to power, providing him with ideological allies in Latin America who espouse similar opposition to U.S. policies in Latin America.

Similar to promises made by Chavez to Venezuelans, Bolivia's president, Juan Evo Morales Ayma, and Ecuador's president, Rafael Vicente Correa Delgado, promised to expand social programs and redistribute incomes during their campaigns. Since coming to power, they have followed similar social and economic policies, although they do not have Venezuela's oil wealth. All three presidents have pushed through new constitutions that grant extensive powers to the president, weakening democratic checks and balances.

Venezuela has sought to influence its neighbors by selling oil at discounted prices. The Venezuelan government has financed the construction of oil refineries in Nicaragua, Ecuador, and Brazil; the expansion of refineries in Uruguay and Paraguay; and infrastructure projects in the Dominican Republic, Jamaica, and Dominica. The Petrocaribe arrangement provides for concessional sales of Venezuelan crude to Caribbean states. Chavez has even donated heating oil to poor families in various areas in the United States under a program started by Joseph Kennedy II. The Venezuelan government has also sought to curry favor with countries by purchasing bonds paying less than market interest rates: It has purchased $500 million of Argentinean bonds, $100 million of Bolivian bonds, and $25 million of Ecuadorian bonds. With Argentinean and Cuban support, Venezuela established La Nueva Televisión del Sur (TeleSUR), a government-controlled network broadcasting the Venezuelan government's spin on events throughout the region.

Cuba. One of the Chavez government's most important international relationships is with Cuba. Chavez considers Fidel Castro his mentor and hopes to inherit the latter's role as the leading figure in the Latin American Revolutionary Left when Castro dies.

Oil is at the center of Venezuela's relationship with Cuba. Venezuela provides more than 90,000 bpd of oil products to Cuba in exchange for the services of some

20,000 Cubans working in Venezuela—teachers, physicians, nurses, sports trainers, military and security advisers, and Chavez's personal security detail.[3] Cuba consumes about 40,000 bpd of this oil; it sells the rest for convertible currency. The value of these resales is quite sizable: At $124 per barrel, the average price for Venezuelan oil in July 2008, 50,000 bpd would yield annual revenues to Cuba of $2.3 billion. Venezuela has also advanced loans to Cuba: Cuba owed Venezuela more than $8 billion as of 2007, of which more than $7.7 billion consisted of long-term debt or unpaid oil deliveries from 2000 to 2007.

Chavez is seeking to institutionalize the Cuba-Venezuela relationship. The Bolivarian Alternative for the Americas (ALBA), launched in Havana in December 2004, provides a "legal umbrella" for integrating the two countries and for "reciprocal assistance." The terms of the agreement include lifting tariffs and taxes on bilateral trade, granting Venezuelan government investors access to 100-percent ownership of investment property in Cuba, and providing access to Cuban ports for Venezuelan ships. Venezuelan airlines have been given the same treatment as Cuban government airlines, including the right to provide domestic passenger service in Cuba. In April 2005, during Chavez's visit to Cuba, the two governments added a number of provisions to ALBA: Forty-nine agreements were signed, encompassing industrial cooperation, transport, agriculture, and sports. In addition, PDVSA and the Cuban state oil company signed joint venture agreements to expand tanker terminals and to jointly refine oil to sell to other Caribbean countries. PDVSA has agreed to help refurbish the Cienfuegos refinery and to build another refinery in Matanzas near the island's main tanker port. PDVSA has also agreed to undertake oil exploration and production in Cuba's exclusive economic zone in the waters surrounding Cuba.

On a grander scale, Chavez has proposed that Cuba join Venezuela in a Bolivarian suprastate. In a broadcast of his weekly television and radio program, Aló Presidente, from Santa Clara, Cuba, on October 14, 2007, the 40th anniversary of Che Guevara's death, Chavez stated that "Cuba and Venezuela could perfectly constitute in the near future a confederation, two republics in one, two countries in one." Cuba's leaders have not expressed any enthusiasm for this proposal.

Cuba's economic dependence on Venezuelan oil largesse is counterbalanced by the dependence of the Chavez government on Cuban expatriate workers to provide social services for Venezuela's poor, key to maintaining their political support.

Venezuela's Wider International Activities. Venezuela consistently opposes U.S. policy initiatives at the United Nations, in Latin America, and elsewhere. Chavez has looked for support from countries outside Latin America to join him in opposing U.S. initiatives. He has sought to make common cause with Iran by creating a strategic alliance, an initiative he discussed during a May 2001 visit to that country. Venezuela has been one of the only countries to vote against referring the Iranian nuclear issue to

---

[3]   Well-informed Venezuelan source, author interview, Caracas, March 2005.

the UN Security Council. There are also reports that Iran and Hizballah are seeking to get around restrictions on financial transactions by opening bank branches in Venezuela and that Hizballah is using its new base in Venezuela for fundraising and other activities in South America. In addition to supporting Iran, Venezuela has supported Russian and Chinese positions when they run counter to U.S. initiatives in an effort to curry favor with the governments of those two countries. Recently, Russia conducted joint naval exercises.

Military Expenditures. Under Chavez, Venezuela's spending on its military has increased from $1.3 billion in 1999 to $2.6 billion in 2006 (IISS, various years). Increased oil revenues have made this possible. They have also enabled the Chavez government to modernize its military and to do so by turning to suppliers other than the United States. Historically, the United States has been the source of Venezuela's military equipment. However, the United States imposed sanctions on Venezuela in September 2005 because of Venezuela's lack of cooperation in the war on terrorism, curbing Venezuela's access to U.S.-manufactured equipment. The United States also blocked the sale of Spanish trainer aircraft and patrol boats, as well as the sale of Brazilian boats, because the equipment contained U.S. components. The Spanish boat sale went forward after the Spanish company replaced U.S. components with European ones.

Facing these obstacles, Venezuela has turned to Russia and China for arms. Venezuela has bought 24 Sukhoi Su-30 fighters, 53 military helicopters, and 100,000 Kalashnikovs from Russia. It has also contracted with Russian companies to build a plant in Venezuela to manufacture Kalashnikovs and ammunition. In May 2008, Chavez announced that Venezuela would buy Chinese K-8 training aircraft.

Destabilization. The most controversial use of oil revenues by the Chavez government has been the support that it has given movements seeking to destabilize neighboring governments. Documents seized from the laptops of Luis Edgar Devia Silva (Raul Reyes), the second in command of the Revolutionary Armed Forces of Colombia–People's Army (Fuerzas Armadas Revolucionarias de Colombia–Ejército del Pueblo, or FARC), killed in a Colombian army raid, show that the Venezuelan government has collaborated with FARC. The documents, disclosed in a news conference in Bogotá by General Óscar Naranjo, head of the Colombian National Police, included a letter from Commander Luciano Marín Arango (also known as Ivan Márquez), a member of the FARC Secretariat and FARC's apparent go-between with Chavez, in which he writes about "Venezuela's financing to the FARC at USD 300 million." In a document dated February 9, 2008, Márquez passes along Chavez's thanks for a $150,000 gift from the FARC when Chavez was imprisoned from 1992 to 1994 for leading a failed coup. Márquez discusses Chavez's plan to try to persuade Latin American governments to get FARC removed from lists of international terrorist groups.

There are indications that Venezuela has supported subversive groups in Peru. A Peruvian parliamentary committee agreed to request the creation of a task force to delve into Venezuelan ideological infiltration in Peru through the so-called Houses of ALBA, a body sponsored and funded by Venezuela. In 2008, Peruvian authorities arrested nine people alleged to be militants bankrolled by Venezuela.

Limitations. Chavez's desire to consolidate power in his own hands and his pursuit of populist economic policy goals have had negative repercussions for Venezuela and, potentially, for himself. This is most apparent in the oil sector. In December 2002, the Venezuelan labor confederation (Confederación de Trabajadores de Venezuela, or CTV), the country's business federation, and the management of PDVSA called a national strike to protest Chavez's policies and in hopes of toppling his government. The strike failed. After the strike, Chavez dismissed nearly 12,000 of PDVSA's 40,000 employees, among them key management and technical personnel, replacing them with a management team loyal to himself. Since then, Venezuelan oil production and exports have steadily declined. Production of oil and associated liquids from natural gas has fallen from 3.46 mbd in 2000 to 2.67 mbd in 2007. Because of increased domestic consumption and smuggling, exports have fallen even more sharply, from 2.96 mbd to 1.93 mbd over the same period. Widespread corruption at PDVSA and the Energy and Oil Ministry also siphons off revenues that should go to the Venezuelan treasury.

Falling oil production bodes ill for the ability of the Chavez government to maintain spending on social programs and to use oil exports sold at concessionary prices to advance its international goals. Nonoil economic growth is unlikely to pick up the slack because of Chavez's policies of nationalizing Venezuelan enterprises, especially those owned by foreign companies, and forcing renegotiations with foreign oil companies to demand more-favorable terms. Foreign and domestic investors are afraid to invest, fearful of the next nationalization.

Although Chavez and his emulators have sought to increase the power of the presidency through changes in local constitutions, they have faced strong resistance from domestic political forces. Latin America is currently enjoying the longest, most stable period of democratic governments in its history. In light of a stronger basis for democracy, the new leaders face strong popular opposition to abandoning elections. Chavez's attempt to abolish presidential term limits was defeated by a national referendum in December 2007; his term ends in 2012.

Chavez also faces opposition to some of his policies from his political base among the poor. Subsidizing Venezuela's neighbors with oil money has been unpopular. Most of Latin America, including Venezuelans, have long ceased to see Cuban economic policies as policies to be emulated. Policy changes in Bolivia and Ecuador are closer to those pursued by Argentina's former dictator, Juan Perón, than to Castro's, but Chavez's policies are geared more toward state control of the economy.

Chavez's own personality has limited his appeal. He has managed to personally alienate the Spanish and Colombian governments. Relations with Brazil cooled after Chavez encouraged Bolivia's President Morales to nationalize natural-gas concessions held by the Brazilian state oil company Petrobras.

Implications for U.S. Security. Revenues from oil exports have enabled Chavez to pursue a number of policies that run counter to U.S. goals to create stable, peaceful, economically dynamic democracies in Latin America. Within Venezuela, he has attempted to concentrate political power in the presidency, undercutting checks and balances. Higher oil revenues permitted him to provide subsidies to his core constituencies, shoring up domestic political support. He has pursued economic policies, and encouraged other countries to do so as well, that retard economic growth and burden government finances. The decline in oil prices in the second half of 2008 is already straining the budget. The Venezuelan government is in the process of reducing some subsidies and devaluing the currency so as to restore fiscal balance.

Chavez has provided campaign financing for presidential candidates in other countries in Latin America who also oppose U.S. policies. Most of these candidates have won recent elections. He has been an irritant to the United States in international forums, such as the United Nations. He has boosted military spending. He has also provided support for groups, such as FARC, that seek to overthrow neighboring governments.

Despite Venezuela's oil revenues, Chavez is far less of a threat to the achievement of U.S. foreign and security policy goals than Iran. He has not won the respect of his neighbors. Although Venezuelan financial assistance is welcome, it has not bought Chavez influence on political and economic policies; their governments go their own ways. Chavez's dream of creating a Bolivarian state has been ignored. In contrast to Iran, Venezuela does not pose a serious military threat to U.S. allies; its two largest neighbors, Brazil and Colombia, have much more capable militaries. In short, increased oil revenues have given Chavez more freedom to pursue policies antithetical to U.S. interests but have not permitted him to become a serious threat to U.S. national security.

Oil and Terrorism Finance

Although the United States lists both al Qaeda and Hizballah as terrorist groups, the two differ markedly in tactics, size, and financial needs. In addition, money used to proselytize for more-extreme forms of Islam may contribute to terrorism by propagating radical interpretations of Islam. Accordingly, we split the discussion of terrorism financing from oil into three sections: the first focusing on small groups, such as al Qaeda; the second on financing efforts to convert individuals to more-extreme versions of Islam; and the third on larger political groups that employ terrorism as one of several instruments in the pursuit of their goals.

Terrorist Groups

The primary goal of al Qaeda, the world's most famous terrorist group, is to overthrow secular or apostate governments in the Muslim world and replace them with Islamic societies governed by the precepts of the Koran. Al Qaeda's primary weapon to achieve this goal is terrorist attacks. Al Qaeda–sponsored terrorist attacks within the region are designed to destabilize existing governments, eventually leading to their overthrow. Terrorist attacks on the United States or U.S. citizens, facilities, or other assets abroad are intended to drive the United States out of the Muslim world. Similar attacks on U.S. allies are designed to do the same.

Funding for al Qaeda and similar, less-known groups originating in the Middle East and North Africa comes from a variety of sources. Donations from groups and individuals from oil-rich countries in the Middle East have been an important source of funding. Higher oil prices may increase the pool of funds available to donors, as disposable incomes in oil exporters rise along with overall increases in economic output. Oil revenues also end up with charities that support al Qaeda and similar groups. Terrorist groups have long used charitable organizations to raise funds (9/11 Commission, 2004, p. 170). Leaders of these charities may know, may not know, or may choose not to know the destination of these funds. Individuals and charities from the GCC nations have been accused of providing funding for al Qaeda and, more recently, the Taliban.

More recently, especially as groups affiliated with al Qaeda have arisen in Europe, other forms of fundraising have risen in importance. Since September 11, 2001, the United States and virtually every country in the world has passed legislation consistent with new international protocols to monitor potential flows of funds to al Qaeda and similar groups. Islamic charities, in particular, have been under great pressure to account for all their donations, although only a tiny fraction of all Islamic charities has donated funds to terrorist groups.

Since September 11, 2001, terrorist groups have shifted more of their efforts to other sources of funding, including drug trafficking, extortion, and illegitimate business operations (Greenberg, 2004). The total cost of the September 11 attacks was approximately $500,000, a relatively small sum of money (Buchanan, 2006). The attacks were planned and prepared over many months. Participants had to pay for living expenses, travel, accommodations, and equipment. The bombings in Madrid in March 2004 cost a total of $80,000 for the entire operation. The London bombings of July 2005 needed about $15,000 (Whitlock, 2008).

More-recent attacks in Europe have been financed locally. Terrorist cells tend to be small. Members often already hold jobs or take jobs before the attack. Materials are affordable: Those for a major bomb attack may be purchased for less than $1,000. For these groups, assistance from donations is no longer necessary to fund their activities.

Islamic Fundamentalism, Terrorism, and Oil Money

Islamic fundamentalism, including some strains of Wahhabism, the fundamentalist brand of Sunni Islam native to Saudi Arabia, has been one of the leading sources of Islamic radicalism in the past few decades.[4] Most Wahhabis do not condone violence. However, extremist Wahhabis have declared those who do not adhere to their beliefs to be heretics, including Muslims from other interpretations of Islam, such as the Shi'a. According to Wahhabi *takfiris* (those who accuse other Muslims of apostasy), Islamic law justifies the murder of these apostates. Unfortunately, extremists within the movement have often come to dominate religious discourse within the Wahhabi sect. Although not the direct cause of terrorist attacks, such as those on 9/11, fundamentalist Islamic ideologies have been the fountainhead of extremist groups across the Middle East, from Algeria to Iraq and Afghanistan, including al Qaeda.

Wahhabism is an integral part of the Saudi state; an alliance between the followers of Al Wahhab and the Al Saud family led to the unification of the Arabian Peninsula and the establishment of the Saudi kingdom in 1932. The Wahhabi religious establishment and the Saudi royal family have since maintained a mutually beneficial relationship.

The U.S. 9/11 Commission (the National Commission on Terrorist Attacks upon the United States) found no evidence that direct funding from the Saudi government found its way into the hands of al Qaeda. Since 9/11, Saudi Arabia has strengthened its antiterrorist financing regime. This said, Saudi Arabia's oil wealth has financed the spread of Wahhabi religious institutions from Bosnia to Great Britain and from Indonesia to the United States. Many Islamic charities and madrasahs have been directly funded by the Saudi government and influenced by the Wahhabi establishment. These institutions have had a substantial influence on Islamic discourse, an influence that they could not have achieved without financial backing from Saudi Arabia. Without financing from Saudi Arabia, it is unlikely that Wahhabism would have as prominent a role in the Islamic community as it does.

Political Groups That Use Terrorism as a Weapon

Hamas and Hizballah are larger, more politically and socially active groups than al Qaeda, but they also use violence to achieve their goals. These groups have large budgets, on the order of hundreds of millions of dollars per year. These expenditures are used for social programs but also to pay the wages and arm militias and security forces that fall under the umbrella of these groups.

Oil-exporting countries have been important sources of funding for Hizballah and Hamas. Iran has been providing as much as $200 million per year in funding for

---

[4] Wahhabism was founded by Muhammad ibn Abd al-Wahhab, an 18th-century fundamentalist Saudi religious reformer. He condemned religious practices and beliefs, such as the veneration of saints, that had emerged over the centuries and sought to return Islam to its "pure" form as purportedly practiced during the time of the prophet Muhammad.

Hizballah (Wilson, 2004). Hamas obtains financial support on the order of a few hundred million dollars annually from Iran and has received assistance from many of the Arab Persian Gulf states (Katzman, 1995). The rulers or governments of these states provide grants to Hamas or affiliated groups from their national budgets or personal fortunes. Increased oil revenues imply increased availability of discretionary spending in the budget; hence, potentially more funding is available for these groups.

Conclusions

Both Chavez and the Iranian leadership have benefited from increased oil export revenues. Without the sixfold increase in Venezuelan government revenues from oil exports, Chavez would have had to spend less on providing subsidies to his neighbors or on the Venezuelan military. Increased oil revenues have enabled the Iranian government to challenge the U.S. presence in the Middle East more assertively. Iran would have had a much more difficult time ignoring UN sanctions if oil prices had been lower. The extent and speed with which it has pursued its nuclear enrichment program would likely have been slower if it had had fewer funds. The decline in world market oil prices in the second half of 2008 has already made it more difficult for both governments to spend on activities contrary to U.S. interests.

However, oil revenues provide a means, not a motivation. North Korea, a decidedly poor country that produces no oil, has developed nuclear weapons without the revenues available to Iran. It too has been one of the U.S.'s primary national security concerns.

Unfortunately, launching a terrorist attack is cheap. The bombings in London and Madrid cost in the thousands, not millions, of dollars. Because of pressure from governments around the world, al Qaeda and its ilk have found it more difficult to rely on donations for their activities. Consequently, al Qaeda and its affiliates have diversified their funding sources to the countries in which they operate or turned to criminal activities for a larger share of their revenues. The terrorists on whom the United States is most focused on pursing have become much less reliant on donations from individuals and charities in oil-rich states. Increases in oil revenues have no bearing on their ability to finance operations.

Higher revenues from exports of oil have helped Iran finance the activities of Hizballah. They also make it easier for Iran and some of the Persian Gulf states to provide assistance to Hamas. However, both Hamas and Hizballah have deep roots in their societies. Even without outside financing, these movements would exist, although probably not at their current strengths.

CHAPTER FIVE

Incremental Costs for U.S. Forces to Secure the Supply and Transit of Oil from the Persian Gulf

Introduction

This chapter analyzes the thesis that the heavy dependence of the United States on imported oil requires substantial additional military forces to maintain the security of international oil flows for the global market. The cost of those forces, in turn, generates a burden on the U.S. taxpayer.[1]

The relationship between ensuring the security of the production and transit of international oil supplies and the costs to the U.S. government is more complex than the straightforward linkage implied in this thesis. Most importantly, military forces are, to a great extent, multipurpose and fungible. Forces designed primarily for use in one theater or in one scenario can typically be brought to bear in another as well. It is also difficult to distill the genesis of a military operation to a unitary aim. Even Operation Desert Storm, which participating nations regarded as necessary to prevent too much of the world's oil production from falling into the hands of a hostile tyrant, was also fought so that Iraq might not be able to violate Kuwait's sovereignty with impunity.

Because oil is fungible, the geographic origin of U.S. imports is irrelevant. As noted in Chapter Two, it is through increases in the price that a major disruption in oil supplies affects the U.S. economy. If the United States were to secure the protection of the production and transit of oil only from sources in the Western Hemisphere or off the Atlantic Coast of Africa, which, combined, are the origin of more than half of current U.S. imports, it would still be vulnerable to a price spike. Violent disruption of production and transit in the Persian Gulf region would lead to a bidding war for oil produced by Nigeria, Venezuela, and others by major importers that currently depend on the Persian Gulf for the bulk of their supplies.

With this in mind, the analysis that follows examines the implied linkage between ensuring secure production and transit of oil from the Persian Gulf and U.S. expenditures for forces specific to this mission.

---

[1] The concern here is with the cost of defense against oil market–related risks rather than with the potential impacts of the risks themselves, which were the focus of Chapters Two through Four.

Key Issues

A Secure Supply of Oil as a Key National Security Interest

Both for the health of the U.S. economy and for broader national security reasons, the United States has given ensured access to oil a high priority among its foreign and defense policy imperatives. The United States has demonstrated its willingness to go to war to prevent the domination of Persian Gulf oil-producing regions by powers hostile to the United States and its allies. The latest National Defense Strategy highlights the critical importance of strategic access to energy resources and the continuing U.S. commitment to maintaining the flow of those resources to the United States and the world economy.

Although the commitment to guarantee access to Persian Gulf oil was a tenet of U.S. declaratory policy before the oil shocks of the 1970s, it was not until the end of that decade that a U.S. administration openly announced its readiness to intervene militarily in the region to prevent a hostile power from dominating the supply of oil from the Persian Gulf. The Carter Doctrine, announced in President Jimmy Carter's State of the Union address on January 23, 1980, stated that the United States could not tolerate being shut out of that region:

> Let our position be absolutely clear: An attempt by any outside force to gain control of the Persian Gulf region will be regarded as an assault on the vital interests of the United States of America, and such an assault will be repelled by any means necessary, including military force.

This statement had clear linkages to military strategy. In September 1978, the Joint Chiefs of Staff issued its "Review of U.S. Strategy Related to the Middle East and the Persian Gulf," in which it named ensuring "continuous access to petroleum resources" as the primary U.S. priority in the region, along with ensuring the survival of Israel.

The clearest manifestation of the expanding U.S. military commitment to protect access to Middle Eastern oil was the creation in 1979 of the Rapid Deployment Force (RDF), which soon gained full, unified command status as the U.S. Central Command (USCENTCOM). It was originally conceived as a package of forces available for worldwide contingencies, although its focus quickly tilted heavily toward the Persian Gulf region after the Soviet invasion of Afghanistan in December 1979 and the announcement of the Carter Doctrine in response a month later. Over the course of the 1980s, considerable forces were made available to USCENTCOM on a priority basis for the purposes of planning, exercises, and operations, as necessary. For the first fiscal year (FY) after the creation of USCENTCOM in 1983 (FY 1984), for example, major combat forces designated to USCENTCOM contingencies included four Army divisions; a reinforced Marine division and a Marine air wing, plus additional Marine forces; seven tactical fighter wings; two strategic bomber squadrons and associated support aircraft making up the Air Force's Strategic Projection Force; three carrier battle groups; two surface action groups; five maritime patrol air squadrons; and three service headquarters.

President Ronald Reagan, nine months into his first year in office, extended President Jimmy Carter's pledge to cover not just external but intraregional threats to U.S. access. In what came to be called the Reagan Corollary to the Carter Doctrine, he made clear the nature of the threat and the U.S. commitment: "there is no way . . . that we could stand by and see [Saudi Arabia] taken over by anyone that would shut off the oil."

The prominence of Middle Eastern oil in U.S. national defense strategy did not diminish as the end of the Cold War approached. National Security Directive (NSD)–26, issued by President George H. W. Bush on October 2, 1989, stated,

> Access to Persian Gulf oil and the security of key friendly states in the area are vital to U.S. national security. The United States remains committed to defend its vital interests in the region, if necessary and appropriate through the use of U.S. military force, against the Soviet Union or any other regional power with interests inimical to our own.

As it turned out, the declaratory national security policy on the criticality of securing the production and transit of oil was not hollow rhetoric. When Iraq's president, Saddam Hussein, invaded Kuwait and threatened Saudi Arabia, President Bush ordered the Pentagon to begin making plans to protect Saudi oil fields and, three days later, authorized Secretary of Defense Dick Cheney to begin deploying U.S. troops to the region. "Our country now imports nearly half the oil it consumes and could face a major threat to its economic independence" were Saudi Arabia to come under hostile control, President Bush said in his address to the American people on August 8, 1990.

NSDs issued during Operations Desert Shield and Desert Storm reiterated the importance of oil security to U.S. strategy in the Middle East. "U.S. interests in the Persian Gulf are vital to the national security. These interests include access to oil and the security and stability of key friendly states in the region," began NSD-45 of August 20, 1990, the presidential directive outlining U.S. policy in response to the Iraqi invasion of Kuwait. On January 15, 1991, NSD-54 repeated this statement of interests and authorized offensive military action against Iraq. President Bill Clinton's administration also cited the United States' critical interest in access to oil, particularly from the Middle East. "Our paramount national security interest in the Middle East is maintaining the unhindered flow of oil from the Persian Gulf to world markets at stable prices," the 1995 *United States Security Strategy for the Middle East* (DoD, 1995, p. 6) read. Efforts to close the Strait of Hormuz "would be of particular concern, since they would touch directly on the availability of oil on world markets." The first objective listed as a USCENTCOM mission at this time was "to ensure uninterrupted access to regional resources (oil)." The 2001 annual defense report from the secretary of defense to Congress expressed the U.S. defense interest in the Middle East and South Asia to be a region at peace, "where access to strategic natural resources at stable prices is unhindered, where no hostile power is able to exercise de facto hegemony, and where free markets are expanding" (DoD, 2001, p. 15).

Although the Iraq war that began in 2003 arose from a number of precipitating factors, a key concern expressed by the George W. Bush administration was the potential for Saddam Hussein, armed with weapons of mass destruction (WMD), to "seek domination of the entire Middle East [and] take control of a great portion of the world's energy supplies" (Cheney, 2002).

Current Policy

Access to foreign oil remains a top priority driving U.S. strategy and defense policy. The National Defense Strategy issued by Secretary of Defense Robert Gates in June 2008 notes that securing access to energy resources is one of the key components of the overall defense strategy aimed to achieve U.S. national security objectives as laid out in the National Security Strategy:

> The United States requires freedom of action in the global commons and strategic access to important regions of the world to meet our national security needs. The well-being of the global economy is contingent on ready access to energy resources. . . . The United States will continue to foster access to and flow of energy resources vital to the world economy. (DoD, 2008, p. 16)

U.S. efforts to ensure secure access to foreign oil go beyond the Persian Gulf region. Since the 1990s, the United States has deepened its ties—economic, political, and military—with oil-producing states in central Asia, South America, and West Africa. The emergence of oil production and export activity off the West African coast has led to an increase, though still modest, in the activities of U.S. military forces in that area. Assistant Secretary of State Walter Kansteiner noted in 2002, "African oil is of national strategic interest to us, and it will increase and become more important as we go forward." In 2003, General James Jones, Supreme Allied Commander Europe (SACEUR) and commander of U.S. European Command (USEUCOM), said that he expected the carrier battle groups of the future in his area of responsibility to spend "half the time going down the west coast of Africa" rather than in the Mediterranean Sea. In November 2007, the USS *Fort McHenry* launched the Africa Partnership Station initiative, providing a persistent presence in the Gulf of Guinea, conducting joint exercises and training, and building the capacity of partner navies to conduct maritime security operations in the region.

Recent Estimates of the Costs of Protecting the Supply and Transit of Oil

Precisely because the level of U.S. dependence on imported oil is a prominent public policy issue, several studies have sought to quantify the costs to the U.S. government of its military efforts to protect oil. The estimates in these studies have varied from as low as $13 billion to $143 billion per year, in 2009 dollars. This wide range of estimates reflects both the complexity of how U.S. forces are planned and operated and, thus, the difficulty of being very specific in allocating precise costs to this mission. Moreover, in some cases, the results appear to be influenced by the policy position of those addressing the problem. In general, the analyses addressed the cost incurred by USCENTCOM in its area of operations to execute the missions of protecting the maritime transit of oil supplies in the Persian Gulf region and Indian Ocean as well as aiding in the defense of friendly oil-producing governments.

Currently, there is no comprehensive, publicly available U.S. government study of the costs of protecting international oil supplies. A study done by the U.S. General Accounting Office (GAO) in 1991 reported DoD estimates of military spending for southwest Asia (SWA) missions from FYs 1980–1990. Over this 11-year period, the GAO reported $27.2 billion in spending directly or primarily related to SWA missions, plus an additional $240 million for reflagging Kuwaiti oil tankers during the Iran-Iraq War. Military ($30.8 billion) and economic ($28 billion) aid to regional governments was cited as well. Most importantly, $272.6 billion was spent for programs motivated by "the need to develop capabilities that would enable the United States to defend its interests not only in Southwest Asia but also in other regions." The bulk of these funds ($220.3 billion), however, was spent to support forces marked as available for RDF and then USCENTCOM contingency operations. In sum, the GAO reported about $360 billion in defense expenditures to defend SWA between FY 1980 and FY 1990, or $32.7 billion per year in nominal dollars—although the figure would have been higher at the end of the decade than the beginning, owing to the expansion of USCENTCOM during that period. This figure corresponds to approximately $66 billion per year in 2009 dollars during the 1980s, although not all of the costs of SWA missions can be attributed uniquely to oil security.

A follow-on study by the Congressional Research Service (CRS) completed in 1992 concluded that nearly all of the costs enumerated in the GAO study would have been borne even in the absence of oil-security demands. After subtracting these costs from the GAO total, the CRS estimated that only $71 billion in nominal dollars, or approximately $13 billion per year in 2009 dollars, could be attributed to oil-security missions in the 1980s. Greene and Leiby (1993) argue that removing *all* readiness costs for SWA from an estimate of oil-security costs is unreasonable. A number of military activities centered in the Persian Gulf region—most notably, U.S. Air Force (USAF) and U.S. Navy (USN) operating costs that were cited in the U.S. military strategy—

support the security of friendly oil-producing nations and ensure maritime transit of oil. Greene and Leiby assume that roughly one-third of these costs were related to oil security, and they add the U.S.-borne costs of Desert Shield and Desert Storm. Based on these assumptions, they calculate that the United States spent roughly $24 billion per year in 2009 dollars on oil-security efforts in the Middle East between 1980 and 1991.

Around the same time, Ravenal (1991) used USCENTCOM's assigned share of U.S. land divisions (four of 17) as an indicator of the proportion of the total defense effort devoted to defending U.S. interests in the Middle East. This rough cut yields an estimate of $67 billion in 2009 dollars on an annual basis for Middle East defense. To this, Ravenal adds about $16 billion per year in 2009 dollars as the expected cost of future regional wars amortized on an annual basis. Kaufmann and Steinbruner (1991) use Joint Military Net Assessments and annual defense reports to outline force-planning contingencies for various regions of the world and estimate that the budget authority for forces assigned to defend all U.S. interests in the Persian Gulf was about $64.5 billion in FY 1990, or about $110 billion per year in 2009 dollars.

More-recent estimates are summarized in Table 5.1. Delucchi and Murphy (2008) use the estimates of Ravenal (1991) and Kaufmann and Steinbruner (1991) as a starting point for peacetime expenditures but discount their totals by a range of $0 to $20 billion to account for the possibility that some percentage of these expenses are fixed and irreducible overhead. They add an expected average war cost in the region of $15 billion to $25 billion per year. This assumes that the total cost of the 1991 Gulf War and the 2003 Iraq War is about $1 trillion and that this sequence of wars occurs every 50 years, yielding an average annual war cost of $20 billion that they express as a range of $15 billion to $25 billion per year (2009 dollars). They further estimate that oil-security interests account for some 60 to 75 percent of U.S. military interests in the Persian Gulf region and that the U.S. government spends something between $29 billion and $75 billion per year in 2009 dollars to provide oil security in the Middle East.

Table 5.1
Recent Estimates of the Costs of Protecting the Supply and Transit of Oil

Study	Scope of Estimate	Annual Cost Estimate (billions of 2009 dollars)
Delucchi and Murphy (2008)	Middle East defense and expected contingencies specific to oil mission	29–75
Copulos (2003)	Middle East defense and contingencies specific to oil mission, 2003	54
Copulos (2007)	Middle East defense and contingencies specific to oil mission, 2007	143

On the high end of published estimates, Copulos (2003) uses DoD budget numbers for personnel and operations and maintenance (O&M) rated against the share of conventional forces assigned to USCENTCOM, plus unique costs, including special operations, pre-positioned materiel, strategic mobility investments, and contingencies for SWA to estimate that $87.2 billion per year was spent on USCENTCOM operations as of early 2003. He estimates that 50 percent of USCENTCOM personnel and O&M costs are attributable to oil-security missions. Based on this, Copulos calculates that DoD spends $44.4 billion per year ($54 billion in 2009 dollars) to secure Middle East oil supplies. In 2007, he updated this analysis, attributing half of the costs of Operation Iraqi Freedom (OIF) to oil-security interests, raising the annual cost to about $143 billion per year in 2009 dollars (Copulos, 2007).

Costing Forces Specific to Ensuring Security of Supply for Oil

The wide range of estimates in Table 5.1 reflects, in part, the ambiguity of assigning specific forces to specific missions. A common approach is to assess the forces needed to ensure the security of production and transportation of oil from the Persian Gulf, the main producing region; estimate their cost; and declare this to be the bill the U.S. taxpayer is paying through DoD to achieve this goal. This is not the approach presented here. It does not follow that, if we were able to remove the mission of ensuring secure production and transit of oil from the Persian Gulf for the global economy, the military could shed the full complement of required forces and the defense budget could be reduced accordingly. For example, some of the forces that were sent to drive Iraqi forces out of Kuwait during Operation Desert Storm were also integral to war plans of combatant commanders other than USCENTCOM.

That said, if a prominent national defense interest of the United States goes away, some force-structure adjustment will typically follow. Most prominent in this regard was the force-structure adjustment that followed the end of the Cold War. A number of analyses during the 1970s and 1980s estimated that 50 percent or more of U.S. force structure was focused on the defense of Europe. When that mission was no longer an imperative, the United States reduced its forces oriented to Europe (proportionately more Army and USAF than maritime forces) by about 30 percent, not 50 percent. The defense budget was reduced, in real terms, by about 20 percent. This reflects three phenomena:

- Not all forces needed to execute a mission can be eliminated once the need for that specific mission passes. Typically, some forces are earmarked for more than one mission.
- Savings are not proportional to the fraction of combat forces eliminated. There are administrative and overhead expenses in the DoD budget that are relatively

inelastic and do not respond in proportion to the reduction (or expansion) of combat forces.

- Because of the impossibility of pinpointing unambiguously the costs of forces incremental to protecting oil security, estimation of these costs will necessarily be expressed as a range, to take into account the imprecision inherent in the nature of the problem. The analysis that follows was developed to produce a reasonable estimate of the additional U.S. forces and defense spending that arise from the need to protect the international supply and transit of oil from the Persian Gulf.

Because this endeavor has, in the past, yielded widely differing estimates and, by its nature, involves some measure of judgment, two research teams operating independently made the determination using different approaches. One team looked at the structure of U.S. forces from the top down, parceling out the defense budget into increasingly differentiated shares, then summing the costs associated with the shares attributed to oil-security missions. Concurrently, two researchers looked at the structure of U.S. forces from the bottom up, identifying the forces specifically, or primarily, associated with oil-security missions, estimating how many of those were likely to be removed from the force if the mission were eliminated, and estimating the costs associated with fielding and maintaining those forces. In this way, we sought to bound a best estimate of the incremental costs to the United States of protecting the international supply and transit of oil from the Persian Gulf. The result is our best estimate of the savings that the U.S. government would likely realize if it were to entirely drop the mission of ensuring the secure production and transit of oil from the Persian Gulf for the global market and reduce its forces in line with the newly reduced demand for the services they provide. This approach reduces, but does not eliminate, the imprecision inherent in the task.

Incremental Forces to Secure Oil from the Persian Gulf

The protection of the supply and transit of oil forms part of the basis for the U.S. force-planning framework for conventional warfare. Up until the terrorist strikes of 9/11, DoD structured its forces to be able to fight two major theater wars (MTWs), one of which was focused explicitly on the Persian Gulf region. In 2004, the National Military Strategy, putting major combat operations into a broader context, stated the need to conduct two overlapping "swift defeat" campaigns. Even when committed to a limited number of lesser contingencies, the force must be able to "win decisively" in one of the two campaigns.

These two data points—the post–Cold War drawdown and the forces needed for two major combat operations—form the starting point for our first estimate of the incremental cost of the U.S. oil-related defense commitment. The post–Cold War drawdown demonstrates that an altered strategic environment can be expected to lead to an adjustment of forces. It also provides us with empirical data on fiscal savings that might be realized from a reduction in force structure. The major combat operations planning framework provides a starting point to estimate what U.S. forces might be *candidates* for reduction if the United States chose to no longer protect oil production and transit from the Persian Gulf.

The post–Cold War reductions in forces and the correlated reductions in military budgets are shown in Tables 5.2, 5.3, and 5.4. The data provide an empirical indication of the degree to which the defense budget was reduced as a prominent U.S. defense priority was eliminated.

The great bulk of the declines in the budget can be ascribed to the reduction in active-duty forces and a slice of the generating force that supported them. Fully burdening each cut in major active-duty units with the budget reductions realized by their respective services, provides an empirical measure of the practical, fiscal result of reducing forces.

Table 5.2
Service Budgets: Total Obligational Authority

Service	FY 1991 Budget	FY 2000 Budget	Difference
Army	148.96	95.3	−54
USN[a]	162.3	115.48	−47
USAF	145.77	109.1	−37

NOTE: All figures in billions of FY 2009 dollars.
[a] The Department of the Navy budget includes the U.S. Marine Corps (USMC).

Table 5.3
Reduction in Service Budgets per Major Unit, FY 1991 Versus FY 2000

Service	Number in FY 1991 Budget	Number in FY 2000 Budget	Difference	Budget Reduction (procurement-corrected) (billions of FY 2009 dollars)	Potential Savings per Major Unit (billions of FY 2009 dollars)
Army divisions	16	10	6	−57	9.5
USN carriers	15	12	3	−45	14.8
USAF fighters and bombers[a]	1,560	936	624	−23	0.04

NOTE: No cuts were made to major USMC combat formations, although there was a modest reduction in end strength.
[a] Specifically, USAF fighters and bombers in the primary aircraft inventory.

The reality is a bit more complex. The force reductions following the end of the Cold War were measurably less than the stated requirements for the defense of Europe had been. This is, in part, because there were demands on those forces beyond the defense of the central region of Europe and because Congress and succeeding administrations felt it prudent to leave some forces stationed in Europe to anchor U.S. forces' interaction with North Atlantic Treaty Organization (NATO) allies and cope with any unexpected turn of events.

The figures have been adjusted to reflect the fact that the procurement budgets were cut disproportionately. Over the long term, this is not a savings. It represents a deficit that will have to be restored to maintain even the reduced force structure. The effect of this adjustment was marginal. The overall change in the budget reduction due to the procurement holiday was less than 10 percent.

With the understanding that major combat operations can take different shapes, the forces used to invade Iraq in March 2003 comprise as clear an example as we have of the force requirements for such an operation. Table 5.4 shows, in round terms, the major U.S. units that participated in OIF.

The absence of protecting the supply and transit of oil from the Persian Gulf as a military mission would hypothetically make the forces for that particular major combat operation a target for reduction. This would not completely eliminate the forces

Table 5.4
OIF Invasion Force: Major U.S. Combat Units

Service	Unit
Army	Two and two-thirds division-equivalents[a]
USMC	One and one-third division-equivalents[b]
USN	5 carrier strike groups[c]
USAF	344 fighters and bombers

[a] This counts the 3rd Infantry Division, the 101st Air Assault Division, one brigade of the 82nd Airborne Division, and the 173rd Airborne Brigade. It does not count the 4th Infantry Division, which was in theater but did not participate in the invasion.

[b] This counts the 1st Marine Division and the 2nd Marine Expeditionary Brigade.

[c] This counts the strike groups associated with the following aircraft carriers: USS *Theodore Roosevelt*, USS *Harry S Truman*, USS *Kitty Hawk*, USS *Abraham Lincoln*, and USS *Constellation*. Some of these aircraft carriers may also have been supporting Operation Enduring Freedom. It does not count the USS *Nimitz*, which relieved the USS *Abraham Lincoln*.

available to USCENTCOM for other operations, including augmenting forces that USEUCOM could allocate to support the defense of Israel if called on to do so.

Using the savings per unit realized by cuts in major force units after the Cold War in Table 5.3 as a rough guideline, the potential savings from forces in Table 5.4 that can be reviewed for reduction are shown in Table 5.5.

Viewed through this lens, the U.S. government might expect to be able to reduce the defense budget by $124 billion, or roughly 20 percent of the projected FY 2009 national defense budget of $611 billion.

If U.S. stated policy of ensuring a secure production and transit of oil from the Persian Gulf were to be repealed, some of those forces would still be required to fill demands elsewhere. Prudence would dictate that forces be available to deploy to the region in the event of a different sort of crisis. Moreover, the construct of two MTWs was never assumed to be a total war-winning effort simultaneously. There was always the idea that some forces would swing from one operation to another. Prudent planning would therefore retain some forces in the force structure to reduce risk in other mission areas. With these things in mind, what would the response of U.S. defense decisionmakers likely be?

Army

Approximately two and two-thirds divisions were employed in the major combat operation portion of OIF. These forces represent around 25 percent of the active-duty Army. Within its projected force levels, as the Army draws down from Iraq, it would likely be able to carry out a remaining major combat operation, with capacity to spare. In considering only major combat operations, it would appear that, if there were no need to protect friendly oil-producing countries, two and two-thirds Army divisions could be cut without taking on excessive risk.

Table 5.5
Possible Savings Realized by Eliminating Forces Needed to Fight One of Two Major Combat Operations

Unit	OIF Force	Potential Savings per Unit (billions of FY 2009 dollars)	Potential Savings (billions of FY 2009 dollars)
Army division-equivalents	Two and two-thirds	9.47	25.2
USMC division-equivalents	One and one-third	9.47	12.3
USN carrier strike groups	5	14.84	74.2
USAF fighters and bombers	344	0.04	12.7
Total			124.3

Marine Corps

Just as no major USMC units were cut after the end of the Cold War, even the disappearance of one major combat operation is unlikely to result in DoD making major cuts in the USMC. USMC units are flexible, expeditionary in character, and adaptable to a broad variety of crises. For that reason, no major formations were cut at the end of the Cold War. Were the military mission of ensuring the security of production and transit of oil to be eliminated, all three active divisions with associated aircraft and lift would still probably be kept in the force structure.

Navy

Oil security has been an increasingly important USN mission since the end of the Cold War, and an important factor in force sizing. As early as the 1980s, the recognition of the importance of protecting sea-lanes for shipping oil was reflected in a greater naval presence in the Persian Gulf, with most of the ships coming from the Pacific Fleet. After the breakup of the Soviet Union, these deployments continued to grow, despite the reduction in the size of the USN's force structure. In 1995, the USN focus on the Persian Gulf region expanded to the point at which the USN reactivated the 5th Fleet and established its headquarters in Bahrain. Of the three carriers that the USN typically kept forward before Operation Enduring Freedom in Afghanistan, one was usually deployed in the Persian Gulf or Indian Ocean. Were ensuring oil security no longer to be a military mission, the most important reason for maintaining a peacetime presence in the Persian Gulf would disappear. With relief from this imperative, a reduction of two carrier task forces from the force structure would still allow the USN to fulfill its other requirements for peacetime presence in other maritime theaters.[2]

This reduction, less than 20 percent of the total surface fleet, would still allow the USN to respond to one major combat operation at a level close to its response to OIF, since the eased forward presence requirement would allow the fleet to be kept in a higher state of readiness.

Air Forces

The USAF deployed 344 fighters and bombers for OIF. That represents about 25 percent of its current fighter and bomber inventory. If one MTW imperative went away, the remaining 75 percent of the current USAF would be able to cope with a major combat operation, with some capacity remaining for other contingencies. High-performance fighter aircraft play an important but numerically modest role in contingencies other than major combat operations.

---

[2] The Navy has traditionally planned for three aircraft carriers to maintain one deployed forward in peacetime. In a surge, it can reduce the factor to two (one forward, one back), even for a relatively extended period if necessary.

In sum, about $67.5 billion could be saved annually in cuts in force structure alone (Table 5.6). This analysis indicates that, if the mission to protect the supply and transit of oil from the Persian Gulf had not existed, the 2009 national defense budget might plausibly have been smaller by about 11 percent.

A Top-Down Look at the Problem

Another team of RAND analysts working independently used a different analytic approach to estimate the costs of the oil-security mission. That team began with the entire defense budget, parceled it out into increasingly differentiated shares, and then summed the costs associated with the shares attributed to oil-security missions.

The core share represents that portion of the defense budget that is not affected, at least in the first order, by adjustments in active-duty force structure. Table 5.7 shows the breakdown of core expenditures.

We estimate that roughly $147 billion of the FY 2009 $518 billion regular defense budget is core and would be unaffected by a reduction in demand for forces from the combatant commands (COCOMs). This leaves $371 billion as *noncore costs*—i.e., expenditures driven by COCOM demand. To be conservative, we reduced this to $350 billion in demand-driven costs. Part of that $350 billion is attributable to demand from oil security–related missions.

We separated the noncore share of the regular defense budget into COCOM shares in relation to an estimate of the share of demand-driven defense spending associated with each COCOM. These estimates are derived from a study that RAND prepared for the Joint Staff. The results are shown in Table 5.8. USCENTCOM, given ongoing U.S. commitments in the Middle East, receives the largest share of the noncore budget.

With this allocation of resources in hand, RAND researchers estimated the proportion of each COCOM's demand that arises from oil-security missions and separated each COCOM's share of the noncore, regular defense budget into oil-driven and

Table 5.6
Cost of Forces Focused on Protecting the Global Flow of Oil

Unit	Force	Cost per Unit (billions of FY 2009 dollars)	Potential Savings (billions of FY 2009 dollars)
Army division-equivalents	Two and two-thirds	9.47	23.8
USN carrier strike groups	2	14.84	29.7
USAF fighters and bombers	344	0.04	12.7
Total			67.5

Table 5.7
Core Versus Noncore (demand-driven) DoD Budget Costs

Budget Area	Budget (billions of FY 2009 dollars)	Core Share (%)	Core Share (billions of FY 2009 dollars)
Personnel	125	33	41.3
O&M	178	25	44.5
Research, development, test, and evaluation (RDTE)	80	25	20
Procurement	104	25	26
Construction	21	50	10.5
Housing	3	75	2.3
Revolving and management funds	3	75	2.3
Total			146.9

Table 5.8
COCOM Shares of the Noncore DoD Budget

COCOM	Estimated Share of Noncore Regular Budget (%)	Implied Noncore Regular Budget ($ billions)
USCENTCOM	35	122.5
USPACOM	20	70
USEUCOM	15	52.5
USAFRICOM	5	17.5
USSOUTHCOM	5	17.5
USNORTHCOM	5	17.5
USSTRATCOM	10	35
USSOCOM	5	17.5
Total	100	350

non–oil-driven shares. The total, as shown in Table 5.9, is $83 billion, of which USCENTCOM provides the dominant share. This $83 billion represents an estimate of the incremental cost to the annual U.S. defense budget of securing the supply and transit of oil.

Costs of Combat Operations

The cost of combat operations related to oil security over and above the cost of maintaining and equipping forces is also difficult to estimate with certainty. There have

Table 5.9
Energy-Security Costs per COCOM

COCOM	Energy-Security Share of Noncore, Nonwar Budget (%)	Implied Energy-Security Noncore Nonwar Budget ($ billions)
USCENTCOM	50	61
USPACOM	10	7
USEUCOM	10	5
USAFRICOM	30	5
USSOUTHCOM	5	1
USNORTHCOM	0	0
USSTRATCOM	0	0
USSOCOM	25	4
Total		83

been a number of occasions when U.S. forces were dispatched to combat in the Persian Gulf in response to immediate or anticipated disruptions in the oil supply. These included Operation Earnest Will to protect Kuwaiti tankers passing through the Straits of Hormuz, Operations Desert Shield and Desert Storm, and OIF. Absent the need to defend sources of oil, the United States would largely avoid that particular set of military operations and their attendant costs. The incremental costs for military operations are typically much less than the base costs of maintaining and equipping forces. Yet, while the costs of past and current operations are known, the four cited are much different in character and in cost. They provide only limited guidance to estimate the frequency and character of future oil security–related military operations and thereby determine the cost that the nation would forgo in their absence.

Operations Earnest Will and Desert Storm have a clear linkage to oil security. The costs of Earnest Will were marginal. The cost of Desert Storm was about $99 billion in FY 2009 dollars, of which $85 billion was borne by allies. OIF is far more complex, far more expansive in its scope, and far more expensive. Moreover, while oil played a role, the declared purpose of the invasion of Iraq included the prevention of Saddam Hussein's government from developing and using WMD and to disrupt the nexus between Saddam and hostile terrorist groups. Whatever the initial mix of declared motivations for war, the desired end state for Iraq includes a stable, peaceful nation that is able to prevent terrorist groups from being funded from or from operating on its soil and to produce and export its oil to the world market.

There is reason to believe that, in the future, the United States will avoid a conflict of this particularly broad scope and ambition. As the operation in Iraq can be expected to wind down, future oil-related military operations are likely to be highly intermittent and characterized more by crisis response, counterterrorism, training of local security forces, and localized stability operations. An operation of this type is much less costly than the invasion and occupation of Iraq has turned out to be. This level of activity could run to $20 billion per year (the rate of U.S. spending in Afghanistan). The cost should be amortized across conflict-free years in order to project an average annual cost. For example, 12 years of relative peace in the Persian Gulf region separated the beginning of Desert Storm and the beginning of OIF. If we assume that operations running eight years out of 20 are plausible, then the annual average cost that could be avoided, spread over 20 years, would be $8 billion. If there is less conflict, the figure will be lower.

Conclusions

The cost to the U.S. taxpayer of protecting the supply and transit of oil from the Persian Gulf has been a hotly debated question. In the assessment in this chapter, two key conclusions emerge. First, the United States does include the security of oil supplies and global transit of oil as a prominent element in its force planning. Second, this does not mean that all the forces that could be earmarked for an operation to protect oil supplies would be dropped from the force structure were the mission to protect the supply and transit of oil to be eliminated. Many of those forces that are included in planning for this mission are included in plans for defending U.S. interests in other regions as well, and the requirement for at least some of the forces would persist even if the mission to protect the supply and transit of oil were to go away. At the same time, the removal of a key defense imperative, ensuring the supply and the safe global transit of oil, would almost certainly lead to some reduction in active-duty forces to reflect this. Were oil security no longer a consideration, the United States could expect to avoid periodic military operations that respond to threats to the production and global transit of oil.

We took two approaches to estimating the savings that could be realized from force cuts. Guided by the post–Cold War drawdown and the forces needed for an MTW, we estimated that $67.5 billion per year could be saved. Added to this is an estimate of $8 billion–per-year savings from military operations that would be avoided. A top-down look at current U.S. allocation of defense resources indicated that $83 billion could be saved annually through force reductions and about $8 billion from military operations that could be forgone. Those figures represent 12 percent and 15 percent of the U.S. defense budget, respectively. In other words, our analysis indicates that the most likely outcome of the removal of the mission to defend oil supplies and sea lines of communication from the Persian Gulf would be a reduction over time of between 12 and 15 percent of the current U.S. defense budget.

The factors that go into the analysis are complex, so the range represents a plausible level that should not be accorded too high a degree of precision. Moreover, our analysis addresses the incremental cost to the defense budget of defending the production and transit of oil. It does not argue that a *partial* reduction of the U.S. dependence on imported oil would yield a proportional reduction in U.S. spending that is focused on this mission. The effect on military cost from such changes in petroleum use would be minimal. That said, the incremental amount the United States spends on its military forces to protect the production and global transit of oil is neither zero nor half of its defense spending—two extreme numbers that have appeared in the debate.

CHAPTER SIX

Policy Options to Address U.S. National Security Concerns Linked to Imported Oil

In the previous chapters, we have critically evaluated links between imported oil and U.S. national security that are commonly suggested by political leaders and commentators. We identified some links between imported oil and U.S. national security but also found that some suggested links are weak or nonexistent. Table 6.1 summarizes our major findings.

In light of these findings, the United States would benefit from policies that diminish the sensitivity of the U.S. economy to an abrupt decline in the supply of oil. The United States would also benefit from policies that would push down the world market price of oil: U.S. terms of trade would improve to the benefit of U.S. consumers, rogue oil exporters would have less money at their disposal, and oil exporters that support Hamas and Hizballah would have less money to give these organizations. The United States might also benefit from more cost-sharing with allies and other nations to protect Persian Gulf oil supplies and transport routes.

Policies that attempt to curtail the likelihood of an oil embargo against the United States or to reduce oil prices to curb terrorist financing are unnecessary or unlikely to

Table 6.1
Potential Links Between Imported Oil and U.S. National Security

Potential Link	Risk or Cost
Large disruption in global supplies of oil	Major
Increases in payments by U.S. consumers due to reductions in supply by oil exporters	Major
Use of energy exports to coerce or influence other countries in ways detrimental to U.S. interests	Minimal
Competition for oil supplies among consuming nations	Minimal
Increased incomes for rogue oil exporters	Moderate
Oil export revenues that finance small terrorist groups	Minimal
Oil export revenues that finance Hamas, Hizballah	Moderate
U.S. budgetary costs of protecting all oil from the Persian Gulf	Moderate

be effective. Oil embargoes have been an ineffective tool for advancing foreign policy goals. Terrorist attacks cost so little that attempting to curtail terrorist financing through measures affecting the oil market will not be effective. Moreover, terrorists are increasingly funded by criminal activities, not donations from oil-exporting states.

In this chapter, we evaluate some commonly suggested policies that might reduce the potential risks to national security from importing oil in terms of effectiveness and cost: economic, political, and environmental. We group these policies into four broad categories:

- policies that mitigate disruptions to the oil supply
- policies to expand domestic sources of supply
- policies to reduce domestic consumption of oil
- policies to reduce U.S. defense expenditures on defending oil supplies from the Persian Gulf.

Adoption of any of these policies involves trade-offs. Our analysis seeks to highlight these trade-offs, thereby informing the debate over how best to structure national energy policy. We conclude with suggestions for a portfolio of energy policies to address U.S. national security concerns linked to imported oil.

Policies to Mitigate Disruptions in the Supply of Oil

Option: Support Well-Functioning Oil Markets

Well-functioning domestic and international petroleum markets are a primary means by which the economic costs of disruptions in the supply of oil can be minimized. Energy prices that are free to adjust to changes in supply and demand, undistorted by subsidies or price controls, offer the most effective mechanism for allocating petroleum in a time of scarcity. The experience of the United States with price controls in the 1970s should be sufficient to dispel any doubt about the wisdom of avoiding that kind of intervention.[1] Subsidy and tax policies that retard long-term responses in increasing energy efficiency and investments in new sources of supply reduce market efficiency, thereby amplifying the potential economic costs of a supply disruption.

Option: Drawing on the Strategic Petroleum Reserve

In 1975, the United States established the SPR to mitigate future supply disruptions in times of national emergency. As of August 2008, the SPR held more than 700 million barrels of oil, equivalent to nearly 60 days of imported oil. The oil is stored at four sites located near major refining centers on the Gulf of Mexico.

---

[1] These experiences included long lines and frequent tank-topping, which wasted time, plus legal prohibitions on reallocating supplies geographically from areas with fuel to areas where it was in shorter supply.

To date, a large-scale drawdown of the reserve has not occurred. However, on a number of occasions, U.S. oil companies have borrowed small amounts of oil from the reserve to alleviate domestic supply disruptions. These loans are reviewed on a case-by-case basis. Companies that utilize the reserve are required to return the oil with additional oil as interest.

As noted in Chapter Two, the United States is not alone in stockpiling oil. All the members of the IEA have committed to maintaining stocks owned by or on which their governments can call in times of emergency. In 2007, more than 1.5 billion barrels, including the 700 million barrels maintained in the SPR, were being held in strategic inventories by OECD countries. This is equivalent to a little more than 30 days of OECD consumption. China has also been building a strategic reserve, reflecting its own concerns about the economic costs of disruptions in oil supplies.

To study the ability of the SPR and international reserves to augment oil supplies following disruptions, thereby mitigating the impact of higher oil prices on the U.S. and global economies, the GAO (2006) developed six hypothetical oil supply–disruption scenarios, ranging from the effects of a hurricane along the U.S. Gulf Coast similar to Hurricanes Katrina and Rita, to a halt in Iranian exports for 18 months, to a catastrophic loss of oil production in Saudi Arabia, resulting in the elimination of Saudi exports for 18 months. The GAO then investigated the potential economic impacts for each scenario under three alternative policies: (1) not releasing oil from the SPR, (2) using only the SPR, and (3) using the SPR in coordination with stocks held by other countries.[2] The effects were evaluated, taking into account the assumed duration of disturbance and daily drawdown capacity, as well as stock sizes.

Table 6.2 shows the projected increases in oil prices (quarterly averages) for the six scenarios. The table clearly shows that small to medium-sized shocks can be mitigated by using the SPR even if other countries do not actively release their strategic stocks. With international coordination, the ability to mitigate disruptions is even larger. The SPR, especially if used in conjunction with other countries' reserves, greatly reduces the threat to the U.S. economy from disruptions in supplies.

However, U.S. policy for use of the SPR is ambiguous, reducing its efficacy. Currently, the SPR can be used only after a presidential declaration of a *national emergency*, which is left undefined. Policymakers have been reluctant to spell out in advance what would trigger SPR use, since, under current law, this would mean defining in advance what constitutes a national emergency related to oil supply disruptions and what responses would be taken.[3]

---

[2] The results presented in GAO (2006) were derived from economic models developed by ORNL and EIA. See Appendix II of GAO (2006) for more on these models.

[3] The cost of ambiguity concerning when to use the SPR is unclear. If a war or accident causes a disruption, gaming may occur if oil-market participants increase inventories in hopes of reaping capital gains later, once the SPR drawdown tapers off. However, the risk of a U.S. drawdown policy being gamed by private-sector participants seems small. Storing oil in quantity is costly—especially for the private sector, which uses aboveground

Table 6.2
Maximum Increase in Average Quarterly Oil Price from Potential Disruptions in Supply

Event	Disruption Length (months)	Disruption Size (tbd)[a]	Policy Scenario (2005 dollars per barrel)		
			No Release	SPR Release	SPR and International Release
U.S. Gulf Coast hurricane	6	155	1–2	0	0
Venezuelan strike	5	307	9–13	0–2	0–2
Iran embargo	18	1,478	19–28	11–17	6–11
Saudi terrorism (lengthy disruption in Saudi production)	8	882	39–67	18–39	15–35
Strait of Hormuz closure	3	882	54–82	32–52	11–24
Saudi shutdown	24	6,205	66–104	60–96	54–87

SOURCE: GAO (2006).

[a] tbd = thousands of barrels per day.

There is a legitimate concern that, if policymakers were to use the SPR reflexively to dampen any sudden jumps in oil prices, private-sector incentives for building inventories and taking other steps to insulate industry from price shocks would be undercut. The absence of a publicly stated policy on when the SPR will be used has the potential to trigger panic hoarding in the event of a significant supply disruption, exacerbating the very conditions that SPR use is supposed to ameliorate. By issuing a public statement that the SPR will definitely be used in the event of a major disruption in supply, the market would be better informed and likely act more temperately if such an event came to pass.

Option: Improving the Resiliency of the Domestic Supply Chain

The economic costs to oil companies of disruptions in supplies because of breakdowns in refining capacity or pipeline operations are potentially large. These concerns provide incentives to private-sector oil companies to make investments in ensuring a robust supply chain. The U.S. government does have a role in ensuring that the supply chain is robust by putting in place permitting processes for constructing and operating refineries, pipelines, and other infrastructure that are transparent and predictable. Problems

storage tanks. Gaming SPR release could be deterred through international cooperation to stop speculative stock building.

with permitting and siting of new facilities could impose a constraint on ensuring the continued resilience of the supply chain.

Ensuring adequate domestic production facilities is not the only solution to forestalling disruptions to the domestic supply chain. If the domestic refining industry has been disrupted, the United States can increase imports of refined oil products. The greater challenge can be obtaining the right kinds of refined oil products on short notice to satisfy local product-quality requirements. U.S. pipeline operators are able to rapidly repair breaks, preventing disruptions from lasting long enough to impose significant economic harm. Refined oil products can be transported by truck, rail, or barge as well as through pipelines, filling any shortfalls in supply.

Disruptions, such as refinery fires, can lead to shortages of reformulated gasoline mandated by state and local environmental authorities to meet federal air-quality standards. Legislation or rules can be crafted to adjust these rules in the event of a local supply disruption until such time as alternative sources of the reformulated fuels become available again.

Policies to Expand Domestic Sources of Supply

Any measures that increase the long-term global supply of refined oil products or close substitutes will increase supply and reduce the market power of oil-exporting countries, thereby lowering the world market price of oil. Lower oil prices not only benefit U.S. consumers; they also reduce incomes for rogue oil exporters and potentially contributions to such organizations as Hamas and Hizballah, thereby enhancing U.S. national security.

The *net* economic benefits of policies designed to reduce world market oil prices by increasing supply will depend on how costly the new or alternative energy sources are relative to oil and on the degree of market power exercised by oil-exporting countries. If the new or alternative sources are cost-competitive, including any spillover effects, such as environmental impacts, then there could be a net economic gain. If the alternative supplies are more costly than refined oil products, the additional costs have to be carefully weighed against the expected benefits. Although the additional supply will put downward pressure on world market prices, as discussed in this section, the decline may be small. Under certain conditions, the potential for reducing the oil import premium that reflects the exercise of exporter market power might justify modest government support to help new alternative fuel technologies surmount technological hurdles (Leiby, 2007; Bartis, Camm, and Ortiz, 2008).

Because of the long lead times involved in expanding supplies either through the development of new oil fields or constructing plants to produce synthetic fuels, the supply-side measures examined in this section will take at least 10 years to bring significant supplies of new fuels to the market. However, prices can respond sooner as market participants factor in new sources of supply.

Option: Open Access to Environmentally Sensitive and Other Restricted Areas

Federal, state, and local governments have restricted oil companies from drilling in some environmentally sensitive areas. The Arctic National Wildlife Refuge (ANWR) in Alaska and the Outer Continental Shelf (OCS) off both the east and west coasts of the United States are two such areas. What would be the likely consequences of opening these areas to exploration and production?

ANWR. A recent study released by the EIA (2008a) concludes that, if ANWR were to be opened up for oil and natural-gas drilling in the near future, oil production would begin in approximately 10 years. During its peak production years in 2025 to 2030, ANWR would likely be capable of providing between 0.5 and 1.5 mbd of production; after 2030, output would be expected to decline as the field would begin to be depleted. The oil contained in ANWR has an estimated value of $374 billion in constant 2005 dollars but would cost approximately $123 billion to extract and bring to market, according to Kotchen and Burger (2007). Of the $251 billion difference, they estimate that approximately $90 billion would benefit the petroleum industry as profit while the remaining $161 billion would flow to the state and federal governments as tax revenue.

OCS. Increases in the price of oil in the first part of 2008 spurred calls to relax or eliminate restrictions on oil exploration and drilling in both ANWR and the OCS. In October 2008, the president and Congress repealed the moratoriums on leasing in the OCS for the purposes of oil and natural-gas drilling. The moratoriums were due to expire in 2012. The EIA (2007a) suggested that, if the ban remains in place until 2012 as had been planned, oil extracted from reserves in the OCS would come online in 2017 and ramp up to a peak extraction rate of 0.2 mbd in 2025. Vidas and Hugman (2008) suggest the potential for considerably greater production rates from the OCS, as high as 0.9 mbd at its peak.

At their peak, expanded access to ANWR and the OCS might add to global supply an amount equal to roughly 4 to 11 percent, with the most likely amount around 7 or 8 percent of baseline U.S. demand. Decisions to develop and produce oil in such areas as ANWR and the OCS involve weighing the value of increasing domestic supplies of oil and associated benefits against the potential environmental risks.

Option: Increase Supplies of Unconventional Fossil Fuels

The development of commercial fossil-fuel substitutes for conventional oil has the potential to reduce demand for imported oil. Unconventional liquid fossil fuels can be produced from coal, oil shale, oil sands, and stranded natural gas. Using estimates compiled by the Task Force on Strategic Unconventional Fuels, the United States is endowed with solid and liquid fuel resources equivalent to approximately 9 trillion barrels of oil, or close to 1,000 years of consumption at current levels (Task Force on Strategic Unconventional Fuels et al., 2006). This is in addition to U.S. Geological Survey (USGS) estimates of the world's recoverable conventional oil resources of more than 3 trillion barrels. The quantity and location of these resources are sizable and relatively well understood. However, the processes, recovery rates, and associated costs required to develop unconventional fuels on a commercially viable scale or the environmental impact (such as wildlife habitat destruction, water and conventional air pollution, and carbon dioxide emissions) that might be caused by greater utilization of these resources are not. These could be significant.

With the exception of Canadian oil sands, current production of unconventional liquids is small. Several sources indicate significant potential for expansion over the longer term. Canada's National Energy Board (NEB, 2007) and EIA (2008e) have developed estimates for increases in fuels produced from oil sands ranging from about 3 to 8 mbd, if oil prices are more than $91 per barrel in 2006 dollars, EIA's assumption in its high-price scenario. If oil prices run closer to the EIA's reference-case scenario, the industry is more likely to produce 1 to 2 mbd. Toman, Griffin, and Lempert (2008) calculate that significant increases in oil sands output would be economically sustainable for oil prices at or above $50 per barrel. The availability of water for some forms of oil sands processing, the environmental effects on water and land, and higher emissions of carbon dioxide than conventional petroleum because of the substantial amounts of energy needed to extract this resource are costs associated with increased production.

Bartis, Camm, and Ortiz (2008) conclude that domestic production of coal-to-liquid (CTL) fuels in the United States could reach 2 to 3 mbd by 2025. The economic competitiveness of CTL is more sensitive to the evolution of technology and especially to the costs of controlling—or penalties for releasing—carbon dioxide to the atmosphere than oil sands. CTL is about twice as carbon dioxide–intensive as conventional oil on a well-to-wheels basis. At oil prices greater than $70 per barrel, significant CTL production growth in the United States may well be economically viable. If carbon dioxide produced during the production of CTL could be captured and stored, motor-vehicle fuels produced through CTL would emit no more carbon dioxide than petroleum-based products. However, the technology for commercial-scale carbon sequestration has not yet been developed. Sequestration will likely be expensive. If biomass were to be mixed with coal and added to the fuel stream, fuels produced through CTL would emit less carbon dioxide than refined oil products.

Gas-to-liquid (GTL) plants convert natural gas into a high-quality, low-sulfur diesel. Because of the value of natural gas for heating and industrial purposes, the economics of GTL plants are not as attractive as CTL unless the gas is *stranded*—i.e., not readily usable for other purposes. When gas fields are connected to larger markets, it may be more expensive than oil and, of course, coal on the basis of comparable energy.

In short, output from Canadian oil sands and U.S. CTL could be enough to replace 15 percent or more of baseline domestic U.S. demand by 2030. The potential for oil shale is more uncertain at this stage, given the need for further advances in technology for extracting fuel from this resource (Bartis, Camm, and Ortiz, 2008).

Option: Increase Supplies of Renewable Fuels (Biofuels)

In April 2007, the U.S. Environmental Protection Agency (EPA) proposed the United States' first Renewable Fuels Standard (RFS). The RFS program mandates an increase in the volume of renewable fuel to be blended into gasoline from 9 billion gallons in 2008 to 36 billion gallons by 2022. The definition of *renewable fuel* in the RFS is not limited to any one particular type. At present, ethanol produced from corn and blended into gasoline is the most widely used renewable fuel in the United States. Biodiesel (produced from soybeans and used cooking oils) also qualifies under the RFS but is used to a much lesser extent.

Ethanol is likely to continue to be the dominant renewable fuel (EPA, 2006). However, substantial additional growth in the output of ethanol would have to come from noncrop, cellulosic feedstocks (corn stalks, brush, and other woody materials). Corn-based ethanol is commercially competitive only because of a $0.51-per-gallon government subsidy. The net energy value of corn-based ethanol, after calculating the energy used to farm, fertilize, and transport the corn crop, is modest and, in some instances, negative. Alternatively, CTL plants could use biomass (Bartis, Camm, and Ortiz, 2008). Foreign suppliers, such as Brazil, could also supply ethanol. Currently, U.S. subsidies for domestic ethanol production and tariffs on foreign-produced ethanol make imports uneconomical.

The U.S. production potential for renewable fuels remains uncertain and subject to debate, although it seems unlikely that it will surpass 10 to 15 percent of U.S. oil consumption in the coming decades. Growing biomass feedstocks and processing them into biofuels will use large amounts of land and water. The impact on global food prices of using land and food crops for motor-vehicle fuels rather than for food was already felt in 2008. Soaring food prices led to a sharp deterioration in the standards of living of the urban poor in developing countries, precipitating food riots in many.

Toman, Griffin, and Lempert (2008) assess the potential range of production costs for U.S. biofuels in 2025 as part of a study analyzing the impacts of imposing a requirement that 25 percent of motor-vehicle fuels and electric-power generation come from renewable fuels. While very favorable technology development could imply costs not much different from baseline costs for petroleum fuels, in less favorable cases, the increase in unit fuel cost could be well in excess of 200 percent. The study finds that biomass is more efficiently used as a fuel to generate electric power than for motor-vehicle fuels.

Policies to Reduce Domestic Consumption of Oil

Like increases in supply, reductions in domestic petroleum demand put downward pressure on oil prices. However, whereas supply-side policies serve to increase the quantity of liquid fuels produced and consumed, policies directed at reducing U.S. consumption of oil would work to make the U.S. economy more energy efficient. Greater efficiency reduces the United States' vulnerability to price shocks because less oil and oil substitutes are needed to generate the same economic output. Diversifying sources of supply also reduces the potential economic costs of a supply disruption. However, by lowering long-term prices, they do not encourage more-efficient use of oil. Like supply-side measures, policies that discourage consumption can take a long time to have a substantial effect on demand because, in many cases, large investments are needed to improve energy efficiency.

Option: Higher Fuel Taxes

A domestic tax on all petroleum consumption will result in higher prices for individual consumers but lower net import payments for the country as a whole, since the world oil price will be lower and the demand for imports should fall. Raising fuel taxes is the most direct way of curbing U.S. consumption of oil. Less consumption would put downward pressure on world market oil prices, potentially reducing some of the national security costs linked to U.S. consumption of imported oil. Higher taxes on oil and refined oil products would also help reduce traffic congestion and accidents, diminish local pollution, and delay global warming.

The effects of U.S. fuel taxes on the world market price for oil will depend heavily on the elasticities of demand for and supply of oil and close substitutes and on the reactions of major oil-exporting countries. The overall economic effects of fuel taxes will depend on the balance between the reduced payments for oil imports and the economic efficiency losses in the U.S. economy from restricting fuel uses.

Fuel taxes have been politically unpopular in the United States even though the United States has the lowest fuel taxes of any industrial country (IEA, 2000).[4] How tax revenues from increased fuel taxes would be used would affect their overall economic impact and probably political opposition to them as well. For example, a per capita refund of revenues from fuel taxes through the U.S. income-tax system would ameliorate some opposition. Revenues could also be used to finance research on alternatives to oil and improving energy efficiency. They could also be used to finance investments

---

[4]   In the United States, the federal government currently levies an $0.184-per-gallon tax on gasoline purchases. States also collect a fuel tax that averages $0.22 per gallon but varies considerably between states, bringing the average total fuel tax paid by motorists to $0.40 per gallon. The effectiveness of fuel taxes at influencing driving behavior and raising revenue has degraded over time due to a steady increase in vehicle fuel efficiency, which has reduced the tax paid by motorists on a per-mile basis and an unwillingness among policymakers to raise the federal tax rate for nearly 15 years. As a result, the fuel tax paid per mile driven has decreased by 40 percent since 1960 (Parry, Walls, and Harrington, 2007) when viewed in real terms.

in infrastructure to improve the efficiency of the U.S. transportation system or permit the use of alternative modes of transportation, such as plug-in hybrid cars.

In addition to their political unpopularity, taxes on oil and refined oil products reduce domestic oil output and profits, although part of the profit lost is recouped in the tax revenues. Fuel taxes may also reduce the price elasticity of U.S. demand, thereby slightly increasing the ability of suppliers to exercise market power.

Option: Policies to Limit Oil Imports

Tariffs or quotas on oil imports affect both domestic supply and domestic demand for oil. Both raise the cost of imported petroleum, tariffs through a tax, and quotas by restricting supply. Higher domestic prices stimulate energy efficiency and conservation, and the drop in global demand causes the world price of oil to decline, all other factors being assumed constant. At the same time, import controls stimulate the production of domestic oil and close substitutes by driving up domestic prices, although these higher domestic prices would have a negative impact on U.S. competitiveness. The net effect of an import-control policy on payments for imports will depend on the stringency of the import limit, the impact on pricing, and the impairment of U.S. competitiveness.

Theoretically, oil tariffs or quotas can be seen as an economically efficient response to a policy problem rooted in the market for oil imports—excessive import costs due to the exporting countries' exercise of market power. Tariffs and quotas can also help reduce long-term vulnerability to oil price shocks by encouraging more-efficient use of petroleum. However, tariffs and quotas have several drawbacks. They require costly monitoring and enforcement. They are economically inefficient because they discriminate against lower-cost foreign producers in favor of higher-cost domestic producers. If domestic production costs are high, quotas, for example, can lead to very high costs for refined oil products, boosting prices for domestic production of oil. The higher costs of refined oil products would make U.S. residents worse off and reduce aggregate U.S. economic output. Moreover, under existing trade agreements, imposing a tariff or quotas on oil may not be permissible; the United States would risk costly retaliatory measures.

Option: Raising Corporate Average Fuel Economy Standards

Congress first implemented the Corporate Average Fuel Economy (CAFE) standard in 1975 following the 1973–1974 Arab oil embargo. The standard requires that automobile manufacturers' sales-weighted average fuel economy meet or exceed a specified minimum standard each year; otherwise, the manufacturer must pay fines based on the number of vehicles sold and the extent to which the standard has been missed. These fines are effectively a tax on the fuel economy of the engine. The standard differs for passenger cars and light trucks, currently 27.5 miles per gallon (mpg) and 22.2 mpg, respectively. Under legislation passed in 2007, standards are slated to increase, and the difference between car and light-truck standards is to be eliminated.

The economic effects of fuel-economy standards are subject to debate. Proponents argue that these policies overcome market barriers faced by consumers, who prefer better fuel economy: Fuel-economy standards induce manufacturers to produce vehicles that are in the long-term economic interest of consumers (Kurani and Turrentine, 2004). Other economists have focused on the costs to manufacturers of producing and selling vehicles when consumers may prefer less fuel-efficient vehicles. Austin and Dinan (2005) have estimated the direct cost of a 10-percent reduction in gasoline consumption through CAFE standards at approximately $3 billion annually. Kleit (2004) estimates that fuel-economy regulation designed to achieve a 7-percent reduction will cost approximately $4 billion annually.[5] Jacobsen (2008) develops an empirical general equilibrium model that he has used to study both CAFE and fuel-tax policy. He finds that increasing gasoline taxes would reduce gasoline consumption for about one-sixth the welfare cost of raising CAFE standards to reach an equivalent drop in consumption. In general, the economic literature indicates that fuel taxes are a more efficient way to reduce consumption of refined oil products than mandates on fuel economy for vehicles.

Policies to Reduce U.S. Expenditures to Defend Oil Supplies from the Persian Gulf

U.S. defense expenditures to safeguard the supply and transit of oil from the Persian Gulf for the global market may run 12 to 15 percent of the FY 2009 U.S. defense budget. Because the international oil market is integrated, even if the United States were to cease importing oil from the Persian Gulf, the United States would still be vulnerable to a disruption in oil flowing from the Persian Gulf because any disruptions in the supply of oil from that area would lead to an increase in the world market price of oil.

In the absence of a repeat of the 1990–1991 Gulf War, other oil-consuming nations are highly unlikely to compensate the United States financially for these expenditures. A more plausible alternative would be to encourage other nations to collaborate in patrolling the sea-lanes and ensuring that oil supplies are secure. The oil-exporting states bordering the Persian Gulf have increased their expenditures on defense in recent years. Not all of these expenditures are desirable from the point of view of the United

---

[5] These estimates ignore any costs and benefits associated with externalities. Kleit (2004) suggests that the carbon and oil dependence benefits from reduced fuel consumption are likely outweighed by additional congestion and accident externalities caused by people driving more due to the reduced operating costs associated with a more fuel-efficient fleet of vehicles. Fischer, Harrington, and Parry (2005) develop an economic model to analyze the effects of increasing CAFE standards that formally accounts for impacts on local pollution, global warming, oil dependence, traffic congestion, and accidents. They find that the magnitude and direction of the welfare change generated by increased fuel-efficiency standards vary across different plausible scenarios.

States—for example, Iran's expenditures. The defense budgets of the United States' traditional European allies continue to be under pressure from demands for spending on pensions and health care. As China and India increase their defense expenditures and capabilities, these two countries may take a greater role in patrolling sea-lanes, including those through the Strait of Hormuz. Whether an increased Chinese naval presence in the Indian Ocean is in the United States' interest is another question.

Policy Effects and Trade-Offs

To illustrate the potential effects on the world oil market of these supply- and demand-side policy measures, we have employed a model developed by Bartis, Camm, and Ortiz (2008) to assess the effects of changes in U.S. supply and demand for oil on world market prices and consumption under varying assumptions about OPEC's supply response to changes in world market prices. Table 6.3 presents the impact on the world market price of oil of reducing U.S. oil imports in 2025 by amounts equal to 10 and 25 percent of total U.S. oil consumption.

We carried out these calculations using the EIA's 2008 high-price case, in which the average imported price of crude oil in 2025 is $91 in constant 2006 dollars. We use the high-price case because prices close to this level are likely to be needed to induce a competitive unconventional fossil or biofuel industry in the United States. In this EIA case, total U.S. consumption of oil and other petroleum-based liquids is about 20 mbd. The ranges for our estimates are a result of using high and low elasticities for global petroleum demand and supply. We also look at two bounding assumptions for an OPEC response to a reduction in demand: one in which OPEC compensates for market pressures for lower prices by cutting supply to support the price, and one in which it accepts the resulting price change with no changes in output.

As would be expected, price reductions grow with the level of reduction in U.S. oil imports. The lower the elasticity, the greater the reduction in price from the drop in U.S. demand for imported oil. Price reductions are also larger when OPEC does not reduce output to shore up prices. In the case with the largest reduction (a reduction in imports equal to 25 percent of U.S. oil consumption), the world market price of oil could fall as little as 2 percent or as much as 12 percent. U.S. oil expenditures would fall due to both a reduction in the amount of oil purchased and a decrease in price paid per barrel. While supply-side policies also have the potential to reduce the price of oil, they are likely to result in an increase in U.S. consumption, which can make their overall impact on oil expenditures ambiguous. Moreover, the supply-side policies are not completely additive: Increased supplies of unconventional fossil fuels or renewables would put downward pressure on prices, discouraging the production of additional supplies of these fuels. This suggests that the most effective policies for reducing the impact of imported oil on U.S. national security are likely to include a combination

Table 6.3
Estimated Reductions in the World Price of Oil Due to Reductions in U.S. Oil Consumption or Increases in the Production of U.S. Fuels

Reduction in Oil Imports	EIA 2025 High Oil Prices (constant 2006 dollars)	
	Low-Effect Case	High-Effect Case
10% of total projected U.S. oil consumption	1–2	2–4
25% of total projected U.S. oil consumption	2–5	5–11

NOTE: EIA *Annual Energy Outlook* (AEO) high oil price for 2025 is $90.90 in constant 2006 dollars. Baseline U.S. demand is about 20 mbd.

of both supply-side policies to diversify sources of supply and demand-side policies to discourage consumption.

Designing Effective Policies for Addressing U.S. Energy-Security Concerns

Table 6.4 summarizes the supply-side and demand-side options presented in the preceding sections. By aggregating various measures, one can develop a portfolio of supply- and demand-side measures that together could reduce U.S. oil imports from the baseline by varying amounts. For example, one could obtain a 10-percent reduction in U.S. demand for imported oil through increases in the production of conventional and unconventional fossil fuels and tax-induced efficiency improvements. Such reductions would provide more diversity of supply, more competition in global fuel markets, lead to lower world market oil prices, and potentially reduce the risk of supply disruptions or other threats to U.S. national security. It should be remembered, however, that all of these policies entail trade-offs. Increased fuel taxes, for example, may reduce the competitiveness of U.S. industry and reduce consumer welfare.

Of these measures, the adoption of the following energy policies by the U.S. government would most effectively reduce the costs to U.S. national security of importing oil:

- Support well-functioning oil markets and refrain from imposing price controls or rationing during times of severe disruptions in supply.
- Initiate a high-level review of prohibitions on exploring and developing new oil fields in restricted areas, in order to provide policymakers and stakeholders with up-to-date and unbiased information on both economic benefits and environmental risks from relaxing those restrictions.

Table 6.4
Summary of Long-Term Supply- and Demand-Side Options

Option	Potential Impact on Long-Term Demand for OPEC Oil	Potential Economic Viability	Key Side Effects & Barriers	Time Frame for Generating Benefits
Increased access to offshore oil & ANWR	Between 4 and 11 percent of baseline U.S. fuel demand	Dependent on specific conditions but likely to be good, given expected oil prices	This has uncertain environmental impacts.	Initial supplies of new oil available in 5–10 years, with peak extraction occurring in 15–20 years after ban is lifted.
Expanded CTL & Canadian oil sands	15 percent or more of baseline U.S. fuel demand	Oil sands likely to be cost-competitive above $50/barrel; CTL, above $70/barrel	High end of demand displacement depends on uncertain prospects for mitigating constraints on oil sands (water) & CTL (carbon dioxide).	Timing is uncertain, but likely on the order of 5–25 years under favorable market conditions.
U.S. oil shale	Uncertain; possibly limited, absent oil prices higher than $70/barrel & further technical advances	Uncertain	Current extraction techniques have large environmental impacts; evolving technologies less.	Timing is uncertain, but likely on the order of 20–30 years under favorable market conditions.
U.S. biofuels	Potentially 10–15 percent of baseline fuel demand	Uncertain; potentially quite costly for more-ambitious targets	Production constrained by limits on biomass availability & cost, trade-offs with food consumption, & environmental impacts, including water consumption.	Timing is uncertain, but likely on the order of 20–30 years, depending on market conditions & breakthroughs on cellulosic ethanol–production technologies.
Fuel tax	Dependent on tax size; relatively moderate taxes could reduce demand up to 15 percent	Rising economic cost with size of tax; relatively moderate taxes unlikely to impose serious economic burdens	Main barrier is political opposition & potentially detrimental effects on U.S. economy from increased fuel costs.	Benefits will grow over time as behavior & investments adjust in response to tax.
Import controls	Dependent on severity of restrictions	Dependent on severity of restrictions; economic impacts similar to but costlier than comparable fuel taxes	International trade legality is uncertain, countervailing exporter action threatened, could reduce competitive pressures in domestic supply markets.	Substitution effect will occur immediately; domestic supply & consumption impacts will occur over the longer term (10–20 years).
Fuel-economy standards	Dependent on stringency of standards	Economic cost likely to exceed that of fuel taxes, especially after recent elevation of CAFE standards	Compromises required to address capacity differences between domestic & foreign vehicle producers & concerns about safety.	Timing depends on timing of new standards; benefits will grow as old vehicle stock is replaced.

- Ensure that licensing and permitting procedures and environmental standards for developing and producing oil and oil substitutes are clear, efficient, balanced in addressing both costs and benefits, and transparent.
- Impose an excise tax on all oil, not just imported oil, to increase fuel economy and soften growth in demand for oil.
- Provide more U.S. government funding for research on improving the efficiency with which the U.S. economy uses oil and competing forms of energy.

Among these, an excise tax on oil is likely to be the most contentious. In order to achieve a 10-percent reduction in demand, the required tax on oil is likely to range between 14 and 33 percent, or $17 to $31 per barrel, assuming a long-run elasticity of demand ranging between −0.3 and −0.7 and a base price of $91 per barrel. Such a tax would plausibly result in a long-run reduction in the world price of oil of between $1 and $4 per barrel, reducing U.S. oil expenditures between $72 billion and $91 billion annually.[6] The revenues raised from the tax would generate on the order of $109 billion to $202 billion per year. If rebated on a per capita basis to U.S. citizens, each U.S. citizen would receive between $326 and $603 per year.

A tax would not be without its costs. The deadweight loss in the U.S. oil market associated with this tax is estimated to range between $6 billion and $11 billion per year. This loss would be borne by U.S. consumers and domestic and foreign oil suppliers operating in the United States. This estimate of deadweight loss should, however, be viewed as an upper bound on the cost of the tax, since oil consumption is associated with a variety of externalities (e.g., traffic congestion, local and global pollution) that would be reduced as a result of the tax.

The policies needed to enhance long-term supply options range from changes in current government policies (in the case of access to previously restricted areas) to more indirect policies (in the case of government-sponsored research and development to increase the economic competitiveness of biofuels). In between these poles, alternative fossil fuels seem poised to expand significantly over the next 20 years without major government intervention in response to market forces. Creating transparent, predictable permitting and leasing procedures to drill and produce oil and manufacture synthetic or renewable fuels will be necessary, if either production of oil from previously restricted areas or synthetic or renewable fuel production is to occur.

The fossil fuel–based options involve large, long-lived capital investments. Before companies make investments to increase fossil-fuel production, policymakers need to address the impact of such investments on goals (set by both President Barack Obama and Senator John McCain during the last presidential campaign) to significantly reduce

---

[6] The reduction in U.S. expenditures is calculated as the sum of a price effect and a quantity effect; the price effect is driven by the change in the world price of oil, while the quantity effect is indicative of the savings obtained from simply consuming 10 percent less oil. In our calculations, the quantity effect tends to dominate the price effect.

greenhouse-gas emissions. They also need to evaluate the impact of these investments on fuel-specific environmental concerns of a more local nature. For renewable fuels, the trade-off is in the capacity to provide energy that has much lower greenhouse-gas emissions than fossil-based alternatives that also is economical, minimizes disruptive effects on food markets, and does not inflict substantial damage on the environment.[7]

In short, importing oil imposes costs affecting the national security of the United States. Developing a more proactive policy framework for the use of the SPR and imposing higher taxes on refined oil products can mitigate these costs. To make higher fuel taxes politically palatable, they would likely need to be refunded to taxpayers, perhaps in the form of rebates through the U.S. income tax. Creating a more predictable, transparent framework for investing in domestic supplies of oil, alternative fuels, and renewables may also be desirable. However, in the case of fossil and renewable fuels, the costs of environmental damage, especially additional emissions of carbon dioxide from synthetic fossil fuels and disruptions to global food markets in the case of renewables, have to be carefully weighed. In light of its past performance, the U.S. government also needs to refrain from trying to pick technological winners. The high cost, low benefits, environmental damage, and disruption to global food markets associated with U.S. ethanol subsidies are only the most recent cases in point.

---

[7] Perhaps an even more challenging long-term trade-off is between alternatives for liquid fuels and a move toward electricity-based systems not dependent on them.

Bibliography

9/11 Commission—*see* National Commission on Terrorist Attacks upon the United States.

Austin, David, and Terry Dinan, "Clearing the Air: The Costs and Consequences of Higher CAFE Standards and Increased Gasoline Taxes," *Journal of Environmental Economics and Management*, Vol. 50, No. 3, November 2005, pp. 562–582.

Balke, Nathan S., Stephen P. A. Brown, and Mine Kuban Yücel, *Oil Price Shocks and the U.S. Economy: Where Does the Asymmetry Originate?* Dallas, Tex.: Federal Reserve Bank of Dallas, 1999.

Barsky, Robert B., and Lutz Kilian, "Oil and the Macroeconomy Since the 1970s," *Journal of Economic Perspectives*, Vol. 18, No. 4, Fall 2004, pp. 115–134.

———, *Do We Really Know That Oil Caused the Great Stagflation? A Monetary Alternative*, Cambridge, Mass.: National Bureau of Economic Research, working paper 8389, July 2001. As of February 25, 2009:
http://www.nber.org/papers/w8389

Bartis, James T., Tom LaTourrette, Lloyd Dixon, D. J. Peterson, and Gary Cecchine, *Oil Shale Development in the United States: Prospects and Policy Issues*, Santa Monica, Calif.: RAND Corporation, MG-414-NETL, 2005. As of February 13, 2009:
http://www.rand.org/pubs/monographs/MG414/

Bartis, James T., Frank Camm, and David S. Ortiz, *Producing Liquid Fuels from Coal: Prospects and Policy Issues*, Santa Monica, Calif.: RAND Corporation, MG-754-AF/NETL, 2008. As of February 13, 2009:
http://www.rand.org/pubs/monographs/MG754/

BEA—*see* Bureau of Economic Analysis.

Beccue, Phillip C., and Hillard G. Huntington, *An Assessment of Oil Market Disruption Risks*, Washington, D.C.: U.S. Department of Energy, EMF SR 8, October 3, 2005. As of February 17, 2009:
http://www.stanford.edu/group/EMF/publications/doc/EMFSR8.pdf

Bernanke, Ben S., Mark Gertler, and Mark Watson, *Systematic Monetary Policy and the Effects of Oil Price Shocks*, New York: C. V. Starr Center for Applied Economics, economic research report 97-25, 1997.

———, "Oil Shocks and Aggregate Macroeconomic Behavior: The Role of Monetary Policy: A Reply," *Journal of Money, Credit, and Banking*, Vol. 36, No. 2, March 2004, pp. 287–291.

BP, *BP Statistical Review of World Energy 2007*, c. June 2007. Update for 2008, as of February 17, 2009:
http://www.bp.com/productlanding.do?categoryId=6929&contentId=7044622

Brown, Stephen P. A., and Mine K. Yücel, "Energy Prices and Aggregate Economic Activity: An Interpretative Survey," *Quarterly Review of Economics and Finance*, Vol. 42, No. 2, 2002, pp. 193–208.

Buchanan, Michael, "London Bombs Cost Just Hundreds," BBC News, January 3, 2006. As of February 17, 2009:
http://news.bbc.co.uk/1/hi/uk/4576346.stm

Bureau of Economic Analysis, "Current-Dollar and 'Real' GDP," spreadsheet, c. January 2008, accessed February 9, 2008. January 2009 version, as of February 17, 2009:
http://bea.gov/national/xls/gdplev.xls

Cheney, Richard B., speech to a national convention of Veterans of Foreign Wars, Nashville, Tenn., August 26, 2002. Excerpts, as of February 25, 2009:
http://query.nytimes.com/gst/fullpage.html?res=9505E3D9113CF934A1575BC0A9649C8B63

Cooper, John C. B., "Price Elasticity of Demand for Crude Oil: Estimates for 23 Countries," *OPEC Review*, Vol. 27, No. 1, March 2003, pp. 1–8.

Copulos, Milton R., *America's Achilles Heel: The Hidden Costs of Imported Oil*, Washington, D.C.: National Defense Council Foundation, October 2003.

———, *The Hidden Cost of Oil: An Update*, Washington, D.C.: National Defense Council Foundation, 2007.

Davis, Paul K., Stuart E. Johnson, Duncan Long, and David C. Gompert, *Developing Resource-Informed Strategic Assessments and Recommendations*, Santa Monica, Calif.: RAND Corporation, MG-703-JS, 2008. As of February 23, 2009:
http://www.rand.org/pubs/monographs/MG703/

Delucchi, Mark A., and James J. Murphy, "US Military Expenditures to Protect the Use of Persian Gulf Oil for Motor Vehicles," *Energy Policy*, Vol. 36, No. 6, June 2008.

DoD—*see* U.S. Department of Defense.

DOE—*see* U.S. Department of Energy.

Downs, Erica Strecker, *China's Quest for Energy Security*, Santa Monica, Calif.: RAND Corporation, MR-1244-AF, 2005. As of February 17, 2009:
http://www.rand.org/pubs/monograph_reports/MR1244/

EIA—*see* Energy Information Administration.

Eller, Stacy L., Peter Hartley, and Kenneth B. Medlock III, *Empirical Evidence on the Operational Efficiency of National Oil Companies*, Houston, Tex.: James A. Baker III Institute for Public Policy, Rice University, March 2007. As of March 3, 2009:
http://www.rice.edu/energy/publications/docs/NOCs/Papers/NOC_Empirical.pdf

Energy Information Administration, "International Petroleum (Oil) Production Data," undated Web page. As of December 6, 2008:
http://www.eia.doe.gov/emeu/international/oilproduction.html

———, *Annual Energy Review 2001*, Washington, D.C.: U.S. Department of Energy, Energy Information Administration, Office of Energy Markets and End Use, November 1, 2002. As of February 13, 2009:
http://tonto.eia.doe.gov/FTPROOT/multifuel/038401.pdf

———, *Annual Energy Outlook 2007*, February 2007a. As of February 13, 2009:
http://www.eia.doe.gov/oiaf/archive/aeo07/

———, *Russia*, country analysis brief, Washington, D.C., April 2007b. May 2008 version, as of February 17, 2009:
http://www.eia.doe.gov/emeu/cabs/Russia/Background.html

———, *International Energy Outlook 2007*, DOE/EIA-0484(2007), May 2007c. As of February 13, 2009:
http://tonto.eia.doe.gov/ftproot/forecasting/0484(2007).pdf

———, *Analysis of Crude Oil Production in the Arctic National Wildlife Refuge*, Washington, D.C.: U.S. Department of Energy, Office of Integrated Analysis and Forecasting, SR/OIAF/2008-03, May 2008a. As of February 13, 2009:
http://www.eia.doe.gov/oiaf/servicerpt/anwr/pdf/sroiaf(2008)03.pdf

———, *Annual Energy Outlook 2008*, DOE/EIA-0383(2008), June 2008b. As of February 14, 2009:
http://www.eia.doe.gov/oiaf/archive/aeo08/

———, "Highlights," *International Energy Outlook 2008*, DOE/EIA-0484(2008), June 2008c. As of February 13, 2009:
http://www.eia.doe.gov/oiaf/ieo/highlights.html

———, *International Energy Outlook 2008*, DOE/EIA-0484(2008), June 2008d. As of February 13, 2009:
http://www.eia.doe.gov/oiaf/ieo/

———, "Projections of Liquid Fuels and Other Petroleum Production in Five Cases Tables (1990–2030)," *International Energy Outlook 2008: Appendix G*, DOE/EIA-0484(2008), June 2008e. As of February 13, 2009:
http://www.eia.doe.gov/oiaf/ieo/ieopol.html

———, "Reference Case Projection Tables 1990–2030," *International Energy Outlook 2008: Appendix A*, DOE/EIA-0484(2008), June 2008f. As of August 8, 2008:
http://www.eia.doe.gov/oiaf/ieo/ieorefcase.html

———, "U.S. Product Supplied for Crude Oil and Petroleum Products," July 28, 2008g. As of February 13, 2009:
http://tonto.eia.doe.gov/dnav/pet/pet_cons_psup_dc_nus_mbblpd_a.htm

———, "U.S. Total Crude Oil and Products Imports," July 28, 2008h. As of August 15, 2008:
http://tonto.eia.doe.gov/dnav/pet/pet_move_impcus_a2_nus_ep00_im0_mbblpd_a.htm

———, "International Natural Gas Reserves and Resources Data: Most Recent Estimates," August 27, 2008i. As of August 8, 2008:
http://www.eia.doe.gov/emeu/international/reserves.xls

———, *Annual Energy Outlook 2009 Early Release*, DOE/EIA-0383(2009), December 2008j. As of August 8, 2008:
http://www.eia.doe.gov/oiaf/aeo/aeoref_tab.html

———, "World Oil Balance 2004–2008," *January 2009 International Petroleum Monthly*, February 6, 2009a. As of December 6, 2008:
http://www.eia.doe.gov/emeu/ipsr/t21.xls

———, "Spot Prices for Crude Oil and Petroleum Products," February 11, 2009b. As of February 13, 2009:
http://tonto.eia.doe.gov/dnav/pet/pet_pri_spt_s1_d.htm

EPA—*see* U.S. Environmental Protection Agency.

Eurasia Group, *China's Overseas Investments in Oil and Gas Production*, New York: US-China Economic and Security Review Commission, October 16, 2006. As of February 17, 2009: http://www.uscc.gov/researchpapers/2006/oil_gas.pdf

Evans, Peter C., *Japan*, Washington, D.C.: Brookings Institution, Brookings Foreign Policy Studies Energy Security Series, December 2006. As of February 17, 2009: http://www.brookings.edu/fp/research/energy/2006japan.pdf

Evans-Pritchard, Ambrose, "Middle East War Threat Rattles Oil Markets," *Telegraph*, September 22, 2008. As of February 17, 2009: http://www.telegraph.co.uk/finance/comment/ambroseevans_pritchard/2792604/Middle-East-war-threat-rattles-oil-markets.html

Feiler, Gil, *Economic Relations Between Egypt and the Gulf Oil States, 1967–2000: Petro Wealth and Patterns of Influence*, Brighton, UK, and Portland, Oreg.: Sussex Academic Press, 2003.

Fischer, Carolyn, Winston Harrington, and Ian W. H. Parry, *Economic Impacts of Tightening the Corporate Average Fuel Economy (CAFE) Standards*, Washington, D.C.: U.S. Environmental Protection Agency and the National Highway Traffic Safety Administration, 2005.

———, "Should Corporate Average Fuel Economy (CAFE) Standards Be Tightened?" *Energy Journal*, Vol. 28, 2007a, pp. 1–29.

———, *Should Automobile Fuel Economy Standards Be Tightened?* Washington, D.C.: Resources for the Future, discussion paper 04-53, revised February 2007b. As of February 25, 2009: http://www.rff.org/rff/Documents/RFF-DP-04-53-REV.pdf

GAO—*see* U.S. Government Accountability Office.

Gately, Dermot, "What Oil Exports Levels Should We Expect from OPEC?" *Energy Journal*, Vol. 28, No. 2, 2007.

Gately, Dermot, and Hillard G. Huntington, "The Asymmetric Effects of Changes in Price and Income on Energy and Oil Demand," *Energy Journal*, Vol. 23, No. 1, 2002.

Goodwin, Phil B., Joyce Dargay, and Mark Hanly, "Elasticities of Road Traffic and Fuel Consumption with Respect to Price and Income: A Review," *Transport Reviews*, Vol. 24, No. 3, May 2004, pp. 275–292.

Graham, Daniel J., and Stephen Glaister, "The Demand for Automobile Fuel: A Survey of Elasticities," *Journal of Transport Economics and Policy*, Vol. 36, No. 1, January 1, 2002, pp. 1–25.

Greenberg, Maurice R., *Update on the Global Campaign Against Terrorist Financing: Second Report of an Independent Task Force on Terrorist Financing*, New York: Council on Foreign Relations, June 15, 2004. As of February 17, 2009: http://www.cfr.org/pdf/Revised%5FTerrorist%5FFinancing.pdf

Greene, David Lloyd, and Sanjana Ahmad, *Costs of U.S. Oil Dependence: 2005 Update*, Oak Ridge, Tenn.: Oak Ridge National Laboratory, ORNL/TM-005/45, January 2005. As of February 17, 2009: http://cta.ornl.gov/cta/Publications/Reports/ORNL_TM2005_45.pdf

Greene, David Lloyd, and Paul N. Leiby, *The Social Costs to the U.S. of Monopolization of the World Oil Market, 1972–1991,* Oak Ridge, Tenn.: Oak Ridge National Laboratory, ORNL-6744, March 1993.

Hamilton, James D., "Statistical Evidence on Macroeconomic Effects of Oil Shocks," presentation, Arlington, Va.: Energy Modeling Forum Workshop on Macroeconomic Impacts of Oil Shocks, February 8, 2005. As of February 17, 2009: http://www.stanford.edu/group/EMF/projects/security/hamilton.pdf

Hamilton, James D., and Ana Maria Herrera, "Oil Shocks and Aggregate Macroeconomic Behavior: The Role of Monetary Policy: A Comment," *Journal of Money, Credit, and Banking*, Volume 36, No. 2, March 2004, pp. 265–286.

House, Trevor, "The Roots of Chinese Oil Investment Abroad," *Asia Policy*, No. 5, January, 2008, pp. 141–166. As of February 17, 2009:
http://www.nbr.org/publications/asia_policy/AP5/AP5_Houser.pdf

Huntington, Hillard G., *Inferred Demand and Supply Elasticities from a Comparison of World Oil Models*, Stanford, Calif.: Energy Modeling Forum, Stanford University, EMF WP 11.5, January 1991.

———, "Oil Price Forecasting in the 1980s: What Went Wrong?" *Energy Journal*, Vol. 15, No. 2, 1994.

IEA—*see* International Energy Agency.

IISS—*see* International Institute for Strategic Studies.

IMF—*see* International Monetary Fund.

International Energy Agency, *Energy Prices and Taxes: First Quarter 2000*, Vol. 2000, No. 2, July 2000, pp. 1–460.

———, *World Energy Outlook 2007*, Paris, November 7, 2007. As of February 13, 2009:
http://www.worldenergyoutlook.org/2007.asp

International Institute for Strategic Studies, *The Military Balance*, various years.

International Monetary Fund, "International Financial Statistics," undated Web page. As of February 9, 2008:
http://www.imfstatistics.org/imf/

———, *World Economic Outlook, April 2007: Spillovers and Cycles in the Global Economy*, Washington, D.C., April 2007. As of November 4, 2008:
http://www.imf.org/external/pubs/ft/weo/2007/01/

Jacobsen, Mark R., "Evaluating U.S. Fuel Economy Standards in a Model with Producer and Household Heterogeneity," San Diego, Calif.: University of California, working paper, April 2008. As of February 17, 2009:
http://econ.ucsd.edu/~m3jacobs/Jacobsen_CAFE.pdf

Jaffe, Amy Myers, "The Changing Role of National Oil Companies in International Energy Markets: Introduction and Summary Conclusions," conference presentation, Houston, Tex.: James A. Baker III Institute for Public Policy, Rice University, March 1–2, 2007. As of March 4, 2009:
http://www.rice.edu/energy/publications/docs/NOCs/Presentations/Hou-Jaffe-KeyFindings.pdf

Jones, Donald W., Paul N. Leiby, and Inja K. Paik, "Oil Price Shocks and the Macroeconomy: What Has been Learned Since 1996," *Energy Journal*, Vol. 25, No. 2, 2004.

Jones, James, commander, U.S. European Command, speech, Defense Writers Group, c. late May–early June, 2003.

Kansteiner, Walter, assistant secretary of state for Africa, March 2002.

Katzman, Kenneth, "Hamas's Foreign Benefactors," *Middle East Quarterly*, Vol. 2, No. 2, June 1995. As of February 17, 2009:
http://www.meforum.org/article/251

Kaufmann, William W., and John D. Steinbruner, *Decisions for Defense: Prospects for a New Order*, Washington, D.C.: Brookings Institution, April 27, 1991.

Kemp, Geoffrey, *Iran and Iraq: The Shia Connection, Soft Power, and the Nuclear Factor*, Washington, D.C.: U.S. Institute of Peace, Iraq and Its Neighbors special report 156, November 2005. As of February 17, 2009:
http://purl.access.gpo.gov/GPO/LPS67579

Kleit, Andrew N., "Impacts of Long-Range Increases in the Corporate Average Fuel Economy (CAFE) Standard," *Economic Inquiry*, Vol. 42, No. 2, April 2004, pp. 279–294.

Kotchen, Matthew J., and Nicholas E. Burger, "Should We Drill in the Arctic National Wildlife Refuge? An Economic Perspective," *Energy Policy*, Vol. 31, No. 6, May 2003, pp. 485–489.

KPMG, *Energy Outlook for China*, Hong Kong, 2005. As of February 17, 2009:
http://www.kpmg.com.cn/en/virtual_library/Industrial_markets/Energy_outlook.pdf

Kurani, Ken, and Thomas Turrentine, *Automobile Buyer Decisions About Fuel Economy and Fuel Efficiency*, Davis, Calif.: Institute of Transportation Studies, University of California, ITS-RR-04-31, September 2004. As of February 17, 2009:
http://repositories.cdlib.org/itsdavis/UCD-ITS-RR-04-31/

Lake, Eli, "Iran Threatens Strait of Hormuz: Key Oil Shipping Passage Could Be Blocked for a Month," *New York Sun*, August 5, 2008. As of February 17, 2009:
http://www.nysun.com/foreign/iran-threatens-to-shut-strait-of-hormuz/83142/

Leiby, Paul N., *Estimating the Energy Security Benefits of Reduced U.S. Oil Imports*, Oak Ridge, Tenn.: Oak Ridge National Laboratory, ORNL/TM-2007/028, February 2007. As of February 17, 2009:
http://www.epa.gov/otaq/renewablefuels/ornl-tm-2007-028.pdf

Leiby, Paul N., and David W. Bowman, *Disruption Scenarios and the Avoided Costs Due to SPR Use*, Oak Ridge, Tenn.: Oak Ridge National Laboratory, working paper, January 19, 2006.

Leiby, Paul N., Donald W. Jones, T. Randall Curlee, and Russell Lee, *Oil Imports: An Assessment of Benefits and Costs*, Oak Ridge, Tenn.: Oak Ridge National Laboratory, ORNL-6851, November 1, 1997. As of February 23, 2009:
http://pzl1.ed.ornl.gov/ORNL6851.pdf

Levy, Walter J., *Oil Strategy and Politics, 1941–1981*, Boulder, Colo.: Westview Press, 1982.

Masaki, Hisane, "Oil-Hungry Japan Looks to Other Sources," *Asia Times*, February 21, 2007. As of February 17, 2009:
http://www.atimes.com/atimes/Japan/IB21Dh01.html

Marrese, Michael, and Jan Vanous, *Soviet Subsidization of Trade with Eastern Europe: A Soviet Perspective*, Berkeley, Calif.: Institute of International Studies, University of California, research series 52, 1983.

Mitsumori, Yaeko, "Slick Operators: Japan's Upstream Industry Gets a Crude Awakening," *Japan Inc.*, October 1, 2003.

National Commission on Terrorist Attacks upon the United States, *The 9/11 Commission Report: Final Report of the National Commission on Terrorist Attacks upon the United States*, New York: Norton, 2004. As of February 17, 2009:
http://www.gpoaccess.gov/911/

National Energy Board, *Canada's Energy Future: Reference Case and Scenarios to 2030*, Calgary, Alta., November 2007.

National Petroleum Council, Committee on Global Oil and Gas, *Hard Truths: Facing the Hard Truths About Energy: A Comprehensive View to 2030 of Global Oil and Natural Gas*, Washington, D.C., July 2007. As of February 14, 2009:
http://www.npchardtruthsreport.org/

National Security Directive 26, U.S. Policy Toward the Persian Gulf, October 2, 1989. As of February 17, 2009:
http://www.fas.org/irp/offdocs/nsd/nsd26.pdf

National Security Directive 45, U.S. Policy in Response to the Iraqi Invasion of Kuwait, August 20, 1990. As of February 17, 2009:
http://www.fas.org/irp/offdocs/nsd/nsd_45.htm

National Security Directive 54, Responding to Iraqi Aggression in the Gulf, January 15, 1991. As of February 17, 2009:
http://www.fas.org/irp/offdocs/nsd/nsd_54.htm

NEB—*see* National Energy Board.

Nordhaus, William D., *Who's Afraid of a Big Bad Oil Shock?*, prepared for Brookings Panel on Economic Activity, Special Anniversary Edition, New Haven, Conn.: Yale University, September 2007. As of February 17, 2009:
http://nordhaus.econ.yale.edu/Big_Bad_Oil_Shock_Meeting.pdf

NPC—*see* National Petroleum Council.

NSD—*see* National Security Directive.

Paik, Keun-Wook, Valerie Marcel, Glada Lahn, John V. Mitchell, and Erkin Adylov, *Trends in Asian NOC Investment Abroad*, London: Royal Institute of International Affairs, Chatham House working background paper, March 2007. As of February 17, 2009:
http://www.chathamhouse.org.uk/files/6427_r0307anoc.pdf

Parry, Ian W. H., and Kenneth A. Small, "Does Britain or the United States Have the Right Gasoline Tax?" *American Economic Review*, Vol. 95, No. 4, September 2005, pp. 1276–1289.

Parry, Ian W. H., Margaret Walls, and Winston Harrington, "Automobile Externalities and Policies," *Journal of Economic Literature*, Vol. 45, No. 2, June 2007, pp. 373–399.

Public Law 88-206, Clean Air Act, December 17, 1963.

Public Law 93-438, Energy Reorganization Act, October 11, 1974.

Public Law 94-163, Energy Policy and Conservation Act, December 22, 1975.

Public Law 94-385, Energy Conservation and Production Act, August 1976.

Public Law 95-617, Public Utilities Regulatory Policies Act, November 9, 1978.

Public Law 95-618, Energy Tax Act, November 9, 1978.

Public Law 95-619, National Energy Conservation Policy Act, November 9, 1978.

Public Law 95-620, Power Plant and Industrial Fuel Use Act, November 9, 1978.

Public Law 95-621, Natural Gas Policy Act, November 9, 1978.

Public Law 110-140, Energy Independence and Security Act, December 19, 2007.

Ravenal, Earl C., *Designing Defense for a New World Order: The Military Budget in 1992 and Beyond*, Washington, D.C.: Cato Institute, 1991.

Shell, "Shell Energy Scenarios to 2050," Web page, February 4, 2008. As of February 13, 2009:
http://www.shell.com/home/content/aboutshell/our_strategy/shell_global_scenarios/
shell_energy_scenarios_2050/shell_energy_scenarios_02042008.html

Task Force on Strategic Unconventional Fuels, U.S. Department of Energy, U.S. Department of Defense, and U.S. Department of the Interior, *Development of America's Strategic Unconventional Fuels Resources: Initial Report to the President and the Congress of the United States*, Washington, D.C.: Task Force on Strategic Unconventional Fuels, September 2006. As of February 17, 2009:
http://www.fossil.energy.gov/programs/reserves/npr/publications/sec369h%5Freport%5Fepact.pdf

Teslik, Lee Hudson, "Royal Dutch Shell CEO on the End of 'Easy Oil,'" Council on Foreign Relations, April 7, 2008. As of August 15, 2008:
http://www.cfr.org/publication/15923/

Toman, Michael, James Griffin, and Robert J. Lempert, *Impact on U.S. Energy Expenditures and Greenhouse-Gas Emissions of Increasing Renewable-Energy Use*, Santa Monica, Calif.: RAND Corporation, TR-384-1-EFC, 2008. As of February 13, 2009:
http://www.rand.org/pubs/technical_reports/TR384-1/

UN—*see* United Nations.

United Nations, "UN comtrade," UN Commodity Trade Statistics Database, New York: UN Statistics Division, undated. As of February 17, 2009:
http://comtrade.un.org/

U.S. Department of Defense, *United States Security Strategy for the Middle East*, Washington, D.C.: U.S. Department of Defense Office of International Security Affairs, May 1995.

———, *Annual Report to the President and the Congress*, Washington, D.C., 2001. As of February 17, 2009:
http://www.dod.mil/execsec/adr2001/

———, *National Defense Strategy*, Washington, D.C., June 2008. As of February 17, 2009:
http://purl.access.gpo.gov/GPO/LPS103291

U.S. Department of Energy, *Energy Policy Act 2005: Section 1837: National Security Review of International Energy Requirements*, Washington, D.C., February 2006. As of February 17, 2009:
http://www.pi.energy.gov/documents/EPACT1837FINAL.pdf

U.S. Environmental Protection Agency, *Renewable Fuel Standard Program: Draft Regulatory Impact Analysis*, Ann Arbor, Mich.: U.S. Environmental Protection Agency, Office of Transportation and Air Quality, EPA420-D-06-008, September 2006.

U.S. Geological Survey, *Mineral Commodity Summaries 2007*, Washington, D.C., January 12, 2007. As of February 13, 2009:
http://minerals.usgs.gov/minerals/pubs/mcs/2007/mcs2007.pdf

U.S. Government Accountability Office, *Strategic Petroleum Reserve: Available Oil Can Provide Significant Benefits, but Many Factors Should Influence Future Decisions About Fill, Use, and Expansion—Report to Congressional Requesters*, Washington, D.C., GAO-06-872, August 2006. As of February 17, 2009:
http://purl.access.gpo.gov/GPO/LPS76688

USGS—*see* U.S. Geological Survey.

Vidas, E. Harry, and Robert H. Hugman, *Strengthening Our Economy: The Untapped U.S. Oil and Gas Resources*, Fairfax, Va.: ICF International, 2008. As of March 30, 2009:
http://www.api.org/Newsroom/upload/Access_Study_Final_Report_12_8_08.pdf

Weisbrot, Mark, and Luis Sandoval, *Update: The Venezuelan Economy in the Chávez Years*, Washington, D.C.: Center for Economic and Policy Research, February 2008. As of February 17, 2009:
http://www.cepr.net/documents/publications/venezuela_update_2008_02.pdf

Weiner, Robert J., "Sheep in Wolves' Clothing? Speculators and Price Volatility in Petroleum Futures," *Quarterly Journal of Economics and Finance*, Vol. 42, No. 2, Summer 2002, pp. 391–400.

Whitlock, Craig, "Al Qaeda Masters Terrorism on the Cheap: Financial Dragnet Largely Bypassed," *Washington Post*, August 24, 2008, p. A01. As of February 17, 2009:
http://www.washingtonpost.com/wp-dyn/content/story/2008/06/23/ST2008062302295.html

Wilson, Scott, "Lebanese Wary of a Rising Hezbollah," *Washington Post*, December 20, 2004. As of February 17, 2009:
http://www.washingtonpost.com/wp-dyn/articles/A12336-2004Dec19.html

Xiaojie, Xu, *Chinese NOCs' Overseas Strategies: Background, Comparison and Remarks*, Houston, Tex.: James A. Baker III Institute for Public Policy, Rice University, 2007. As of February 17, 2009:
http://www.rice.edu/energy/publications/docs/NOCs/Papers/NOC_ChineseNOCs_Xu.pdf

Yergin, Daniel, *The Prize: The Epic Quest for Oil, Money, and Power*, New York: Simon and Schuster, 1991.

Ziegler, Charles E., "Competing for Markets and Influence: Asian National Oil Companies in Eurasia," *Asian Perspective*, Vol. 32, No. 1, 2008, pp. 129–163. As of February 17, 2009:
http://www.asianperspective.org/articles/v32n1-e.pdf